AF412500

Models of Exacerbations in Asthma and COPD

....

Contributions to Microbiology

Vol. 14

Series Editors

Axel Schmidt, Witten
Heiko Herwald, Lund

Models of Exacerbations in Asthma and COPD

Volume Editors

Ulf Sjöbring, Lund
John D. Taylor, Lund

12 figures, 4 in color, and 4 tables, 2007

KARGER

Basel · Freiburg · Paris · London · New York ·
Bangalore · Bangkok · Singapore · Tokyo · Sydney

Contributions to Microbiology

formerly 'Concepts in Immunopathology' and
'Contributions to Microbiology and Immunology'

· ·

Dr. U. Sjöbring
Medical Science
Respiratory and Inflammation
Therapeutic Area
AstraZeneca R&D
Scheelevägen 2
SE-221 87 Lund (Sweden)

Dr. J. D. Taylor
Biological Science
Respiratory and Inflammation
Research Area
AstraZeneca R&D
Scheelevägen 2
SE-221 87 Lund (Sweden)

Library of Congress Cataloging-in-Publication Data

Models of exacerbations in asthma and COPD / volume editors, Ulf Sjöbring,
John D. Taylor.
        p. ; cm. – (Contributions to microbiology, ISSN 1420-9519 ; v. 14)
    Includes bibliographical references and indexes.
    ISBN-13 : 978-3-8055-8332-9 (hard cover : alk. paper)
    1.   Asthma–Microbiology. 2.   Asthma–Animal models. 3.   Lungs Diseases,
Obstructive–Microbiology. 4.   Lungs Diseases, Obstructive–Animal models.
    [DNLM: 1.   Asthma–prevention & control. 2.   Asthma–microbiology. 3.
Models, Animal. 4.   Models, Biological. 5.   Pulmonary Disease, Chronic
Obstructive–microbiology. 6.   Pulmonary Disease, Chronic Obstructive–
prevention & control. WF 553 M6887 2007] I.   Sjöbring, Ulf. II.   Taylor, John
D., Dr. III.   Series.
    RC591.M63 2007
    616.2'38–dc22
                                                      2007028107

Bibliographic Indices. This publication is listed in bibliographic services, including Current Contents® and Index Medicus.

© Copyright 2007 by S. Karger AG, P.O. Box, CH–4009 Basel (Switzerland)
www.karger.com
Printed in Switzerland on acid-free and non-aging paper (ISO 9706) by Reinhardt Druck, Basel
ISSN 1420–9519
ISBN 978–3–8055–8332–9

Contents

VIII Foreword
Sjöbring, U.; Taylor, J.D. (Lund)

Introduction

1 Exacerbations of Asthma and COPD: Definitions, Clinical Manifestations and Epidemiology
O'Byrne, P.M. (Hamilton, Ont.)

Human Asthma Models

12 Human Rhinovirus Models in Asthma
Singh, A.M.; Busse, W.W. (Madison, Wisc.)

21 Allergen Inhalation Challenge: A Human Model of Asthma Exacerbation
Gauvreau, G.M.; Evans, M.Y. (Hamilton, Ont.)

Animal Asthma Models

33 Cellular and Animals Models for Rhinovirus Infection in Asthma
Xatzipsalti, M.; Papadopoulos, N.G. (Athens)

42 Modeling Responses to Respiratory House Dust Mite Exposure
Cates, E.C.; Fattouh, R.; Johnson, J.R.; Llop-Guevara, A.; Jordana, M. (Hamilton, Ont.)

68 Respiratory Syncytial Virus-Induced Pulmonary Disease and Exacerbation of Allergic Asthma
Lukacs, N.W.; Smit, J.; Lindell, D.; Schaller, M. (Ann Arbor, Mich.)

Human COPD Models

83 Lipopolysaccharide Challenge of Humans as a Model for Chronic Obstructive Lung Disease Exacerbations
Kharitonov, S.A. (London); Sjöbring, U. (Lund)

101 A Human Rhinovirus Model of Chronic Obstructive Pulmonary Disease Exacerbations
Contoli, M. (Ferrara/London); Caramori, G. (Ferrara); Mallia, P. (London); Papi, A. (Ferrara); Johnston, S.L. (London)

Animal COPD Models

113 Animal Models of Cigarette Smoke-Induced Chronic Obstructive Lung Disease
Churg, A.; Wright, J.L. (Vancouver, B.C.)

126 Animal Models of Chronic Obstructive Pulmonary Disease Exacerbations
Gaschler, G.J.; Bauer, C.M.T.; Zavitz, C.C.J.; Stämpfli, M.R. (Hamilton, Ont.)

142 Author Index

143 Subject Index

This publication was supported by an unrestricted grant from AstraZeneca

Foreword

With the current volume of the Karger book series *Contributions to Microbiology*, we attempt to provide an overview of models of exacerbations of asthma and chronic obstructive pulmonary disease. This is a large area so within one volume we cannot be totally comprehensive. However, we have attempted to point to experimental systems that mimic the pathobiological processes likely to be critical in the exacerbations of these conditions. By the same token we asked the contributors to describe both animal models and human experimental models, since we believe that knowledge generated by linking both these systems will be required to generate insight into the mechanisms of exacerbations of pulmonary disease and the discovery of future treatments. For this update, we were very honored to have been able to recruit some of the most eminent scientists within the area of pulmonology to give their insights into this evolving field of human medicine. We hope that this volume will serve to stimulate and provoke thoughts in a research area where much work remains to be done to meet the high unmet medical needs.

Ulf Sjöbring
John D. Taylor
Lund, May 2007

Sjöbring U, Taylor JD (eds): Models of Exacerbations in Asthma and COPD.
Contrib Microbiol. Basel, Karger, 2007, vol 14, pp 1–11

Exacerbations of Asthma and COPD: Definitions, Clinical Manifestations and Epidemiology

Paul M. O'Byrne

Firestone Institute for Respiratory Health, St. Joseph's Hospital, Hamilton,
Ont., Canada

Abstract

Exacerbations are important events in patients with asthma and chronic obstructive pulmonary disease (COPD). Reducing the number, frequency and the severity of exacerbations is therefore an important management goal identified by treatment guidelines for both diseases. There are similarities with respect to clinical manifestations, etiology and epidemiology, but there are also clear differences not the least in how well patients respond to treatment. The similarities and differences between exacerbations asthma and COPD with an emphasis on epidemiology and clinical presentation are discussed and compared in this chapter.

Introduction

Exacerbations are important events in the lives of patients with asthma and chronic obstructive pulmonary disease (COPD), and reducing their number and frequency is an important management goal identified by treatment guidelines for both diseases [1, 2]. Indeed, it could be argued that reducing the risk of exacerbations is the most important management goal, because exacerbations constitute the greatest risk to patients, are a great cause of anxiety to patients and their families, the greatest stress on health care providers and generate the greatest cost to the health care system in providing management [3]. Surprisingly, despite the importance in preventing either asthma or COPD exacerbations, these have often not been clearly defined in studies and have only relatively recently (at least for asthma) become a major outcome variable in research into the efficacy of drug treatment. Indeed, the first study to be powered to examine an effect of treatment on severe asthma exacerbations was published a little more than 10 years ago [4], while studies of the management and prevention of COPD exacerbations extend back more than 40 years [5].

Definitions of Asthma Exacerbations

A working group has recently been developed on establishing a practical, useful and responsive definition of asthma exacerbations, as part of a larger initiative established by the European Respiratory Society and the American Thoracic Society on establishing recommendations on 'asthma control'. This working group has suggested that severe asthma exacerbations are events, which require urgent action on the part of the patient and physician to prevent a serious outcome, such as hospitalization or death from asthma. The occurrence of severe asthma exacerbations should be used as a marker of poor asthma control.

A definition of a severe asthma exacerbation should include, at a minimum, at least one of the following:

- The need for oral/systemic corticosteroids at the investigator's discretion. The corticosteroids should be administered for at least 3 days, in an effort to avoid including inadvertent and possibly inappropriate use by patients. While this component suffers from lack of precision, it is clinically valid, both sensitive and reasonably specific, and responsive to treatment.
- A hospitalization/emergency room (ER) visit because of asthma requiring systemic corticosteroids. (This definition will exclude those patients attending ER without a severe exacerbation, for example because they have run out of medication.)

This definition is more suited to retrospective diagnosis of severe asthma exacerbations, for example in assessment of clinical trial outcomes, than for use in clinical practice guidelines describing how to treat severe exacerbations, or for prospective guidance for investigators about how to manage exacerbations occurring during the course of a clinical trial, because of the essentially circular argument about the definition of a severe exacerbation (by the use of oral corticosteroids) and the way in which severe exacerbations, once defined, should be treated (with oral corticosteroids).

In the reporting of severe exacerbations, the information on which of the individual criteria had been met by a patient, so that the event had been recognized as an exacerbation, should be provided.

The inclusion of a change in peak expired flows (PEF) from baseline (most often a decline of at least 30% for 2 consecutive days) does not appear to be a sensitive, or a clinically valid, component of a severe exacerbation definition [6]. Other statistical methods for calculating PEF change may be more robust [7], but this requires more information before being included in a definition. Finally, changes in symptoms and/or β_2-agonist use are sometimes used as part of a definition of severe asthma exacerbations, but without the least justification with regard to an important magnitude of change. If included, the changes in PEF, symptoms and/or β_2-agonist use should persist for more than 1 day (unless very severe) to qualify as a severe exacerbation.

Moderate asthma exacerbations are events which, when recognized, should result in a change in treatment, in an effort to prevent the exacerbation becoming severe; therefore, the concept of a moderate exacerbation has clinical utility. The definition of a moderate asthma exacerbation should include deterioration in symptoms, rescue bronchodilator use and/or lung function which lasts at least 2 days, but which is not severe enough to warrant oral corticosteroid use and/or a hospital visit. The precise enumeration of the magnitude of change in these outcomes will differ depending on the population studied and each individual patient's baseline variation. More work is needed to establish definitions, which take into account the patient's usual range of variation.

A definition of a 'mild asthma exacerbation' is not justifiable with present methods of analysis, because these episodes will be only just outside the normal range of variation for the individual patient and may reflect transient loss of asthma control rather than the early stages of a severe exacerbation.

The definitions which may be suitable for clinical trials may not necessarily be suitable for use in clinical practice, because of differences in health care resources (e.g. 24-hour access to investigators in clinical trials), patient and clinician expectations about, and experience with, monitoring of symptoms or PEF, and the overriding need for prospective rather than retrospective definitions to provide guidance for health care professionals in treating exacerbations in clinical practice. In the clinical setting, the absolute severity of exacerbations will vary considerably from patient to patient, or over time. The clinical characteristics, which cause acute distress to one patient, may represent another patient's usual status. Therefore, asthma exacerbations should be *clinically* identified by changes in symptoms and/or rescue use and/or in lung function, which are outside the patient's usual range of day-to-day asthma variation.

Definitions of COPD Exacerbations

Efforts to define a COPD exacerbation began earlier than for asthma. In 2000, Rodriguez-Roisin [8] summarized the proceedings of a workshop to develop a definition of an exacerbation of COPD. The definition of a COPD exacerbation that was agreed upon is 'a sustained worsening of the patient's condition, from the stable state and beyond normal day-to-day variations, that is

Table 1. A scale for COPD exacerbation severity incorporating exacerbations managed at home and in hospital [reproduced with permission from 9]

Mild	An exacerbation treated with antibiotics but no systemic corticosteroid; if no blood gases are available the absence of respiratory failure is assumed
Moderate	An exacerbation treated with parenteral corticosteroids with or without an antibiotic; if no blood gases are available the absence of respiratory failure is assumed
Severe	Type 1 respiratory failure with hypoxia but no carbon dioxide retention or acidosis; P_{aO_2} <8 kPa (60 mm Hg) and P_{aCO_2} <6 kPa (45 mm Hg)
Very severe	Type 2 respiratory failure, compensated with hypoxia, carbon dioxide retention but no acidosis; P_{aO_2} <8 kPa (60 mm Hg), P_{aCO_2} >6 kPa (45 mm Hg) and hydrogen ion concentration <44 nM (pH >7.35)
Life-threatening	Type 2 respiratory failure, decompensated with acidosis and carbon dioxide retention; P_{aCO_2} >6 kPa (45 mm Hg) and hydrogen ion concentration >44 nM (pH <7.35)

P_{aO2} = Arterial oxygen tension; P_{aCO_2} = arterial carbon dioxide tension.

acute in onset and necessitates a change in regular medication in a patient with underlying COPD'. This change in treatment can be qualified to include: (1) requiring treatment with oral or parenteral corticosteroids and/or (2) requiring treatment with an antibiotic. Subsequently, Burge and Wedzicha [9] extended the definition to include a scale of severity of an exacerbation from mild to life-threatening (table 1).

Clinical Manifestations of Asthma Exacerbations

Severe asthma exacerbations are episodes of increased asthma symptoms, such as dyspnea, cough, wheezing and/or chest discomfort. Patients are usually very anxious, hyperventilating, and have a tachycardia. Severe exacerbations are often associated with hypoxic respiratory failure and, when very severe, with hypercapneic respiratory failure. The signs of hypercapneic respiratory failure are the inability to speak in sentences and obtundation. These clinical signs indicate that the patient has a potentially fatal asthma exacerbation and indicate an acute medical emergency, requiring immediate treatment [2].

Asthma exacerbations are associated with (often severe) reductions in airway caliber, usually measured by a fall in the forced expired volume in 1 s (FEV$_1$) or in PEF. Ideally, measurements of FEV$_1$ should always be made when patients are seen in an emergency department setting, both to evaluate the severity of the exacerbation and to monitor the response to treatment.

Asthma exacerbations, in general, develop gradually over 5–7 days before being recognized and treated [10]. Once treatment with oral corticosteroids is started, the exacerbation usually resolves over 5–7 days. However, in occasional patients, severe life-threatening exacerbations can develop over minutes to hours [11]. A recent study has reported on in-hospital asthma mortality in the United States in patients admitted with acute asthma exacerbations [12]. The mortality rate was 0.5% (99% CI 0.4–0.6) and deaths in this population accounted for about one third of all asthma deaths reported in the United States. Interestingly, black patients hospitalized for asthma exacerbations were less likely to die when compared with Caucasian patients; however, this was not explained by race differences in hospital deaths and was attributed to factors preceding hospitalization.

In clinical practice, the concept of a spectrum of severity of exacerbations is well established. Clinicians recognize episodes which are troublesome to patients, and which prompt a desire for a change in treatment (even if only increased reliever use), but which are not severe. In the clinical setting, the absolute severity of such episodes may vary considerably from patient to patient, or over time, in that the clinical characteristics which cause acute distress to one patient may represent another patient's usual status. These events are therefore clinically identified by being outside the patient's usual range of day-to-day asthma variation.

Clinical Manifestations of COPD Exacerbations

COPD exacerbations are characterized by increased dyspnea, cough, sputum volume or purulence, wheeze, and chest tightness. Also, patients can experience fatigue, exercise intolerance, peripheral edema, confusion, cyanosis, and obtundation. Anthonisen et al. [13] identified three cardinal symptoms and used these to classify exacerbations. These symptoms are increased dyspnea, increased sputum volume and sputum purulence. Exacerbations were then graded into type 1 (all three cardinal symptoms), type 2 (two cardinal symptoms), and type 3 (one cardinal symptom plus one of the following: an upper respiratory tract infection in the past 5 days, fever without other cause, increased wheezing or cough, or an increase in heart rate or respiratory rate by 20% compared with baseline readings). The changes in lung function, because most of the airflow obstruction in COPD is irreversible, are much less marked that in asthma. Arterial blood gases and oxygen saturation may also worsen, and severe exacerbations may progress to respiratory failure.

The onset and resolution of a COPD exacerbation has been described by Seemungal et al. [14]. These investigators identified a prodromal phase, during which respiratory symptoms of dyspnea, sore throat, cough, and nasal congestion worsened for 7 days before the onset of exacerbation. On the day the exacerbation was identified, symptoms increased sharply, but the changes in lung function were small. The exacerbation took, on average, 7–14 days to recover. However, a much longer period of up to several months was needed to recover health status as measured by health-related quality of life. The longer duration of worsening health status compared with lung function suggests that exacerbations have a lasting impact.

Some COPD patients experience frequent exacerbations. Not surprisingly, these patients have a poorer quality of life [15], are more likely to be readmitted to hospital [16] and are associated with higher rates of mortality [17]. Several studies have examined the long-term consequences of COPD exacerbations on lung function. These studies have been consistent in their conclusions that patients with frequent COPD exacerbation (>2/year) had a more rapid decline in FEV_1 [18–20].

The mortality rates in patients admitted to hospital with an acute exacerbation of COPD have ranged from 4 to 30%. This variability can in part be explained by the different subgroups of patients that were evaluated in different studies [21]. In patients without respiratory failure, the mortality rates range from 5 to 11%; however, in patients admitted with respiratory failure the mortality rates are higher ranging from 11 to 26%. Also, mortality rates appear to be higher in smaller, nonteaching hospitals when compared to teaching hospitals [22]. Following hospital discharge after treatment for an acute exacerbation, in the United Kingdom, 14% of patients die within 3 months [16]. Similar information is available from the United States, where hospital mortality rates were 11%, 20% after 2 months, 33% after 6 months, and 43% after 12 months [23], and in the Netherlands, where hospital mortality rates were 8%, increasing to 23% after 1 year after discharge [24]. In these studies, common predictive factors for mortality were severity of illness, body mass index, age, prior functional status, abnormalities in gas exchange, a high comorbidity index, and prior hospital admissions.

Epidemiology of Asthma Exacerbations

It is very difficult to accurately estimate the prevalence of even severe asthma exacerbations and comment about the change in prevalence over time. Asthma mortality is a surrogate for the most severe exacerbations and the risk of death from asthma has declined over the last two or three decades in children and adults in both the United States and Britain [25, 26]. Importantly, while a decline in asthma deaths had occurred in the black population in the United States, their risk of asthma death remains over 3 times that of white patients [26]. Also, asthma mortality prevalence changes with age, with asthma deaths being rare in children and increasing to 6.71/100,000 in adults over 65 years [27].

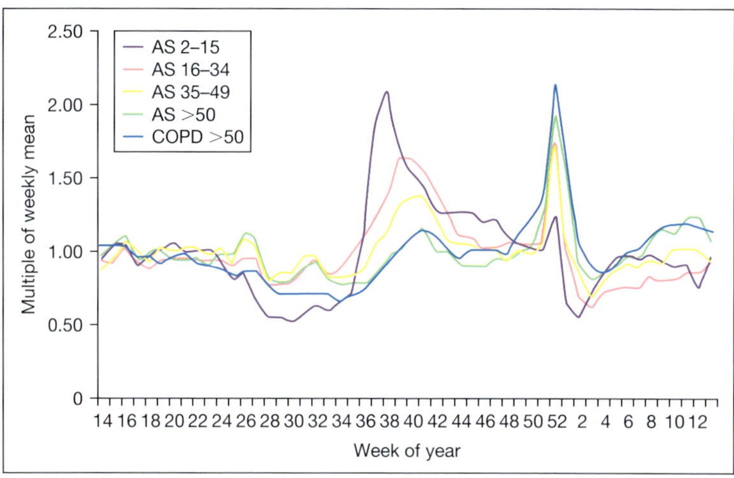

Fig. 1. Annual cycles of ER presentations for asthma (AS) and COPD as multiples of the weekly mean number by week of the year, combining all events from April 2001 to March 2004 in Ontario, Canada. The scale has been set to run from week 14 to week 13 to permit an unbroken view of the new year period [reproduced with permission from Neil W. Johnston, Firestore Institute for Respiratory Health, Hamilton, Ontario, Canada].

Variations in the prevalence of asthma exacerbations are associated with seasons of the year. Seasonal cycles of asthma exacerbations requiring hospital treatment have been reported in many northern hemisphere countries including Canada, the USA, the UK, Mexico, Israel, Finland, and Trinidad, West Indies [28]. Similar cycles have also been reported in Australia and New Zealand. The epidemic peak in early autumn, in all countries examined, is of greatest magnitude in children, less in younger adults, and barely detectable in the elderly (fig. 1). One reason for this is the exposure to environmental allergens, such as those caused by pollens [29] and moulds, such as *Alternaria* [30]. However, environmental allergen exposures do not explain the majority of these severe exacerbations, especially in children. The most likely explanation for a marked increase in hospital visits for severe asthma in the early fall in the northern hemisphere is related to factors that condition the transmission of rhinovirus infection [31, 32]. These include whether schools are open and the occurrence of other events that may foster rhinovirus transmission. While the majority of asthma exacerbations coincide with rhinovirus infections, the effects of these on asthmatics may be amplified by poor air quality (both ambient and domiciliary) and exposure to allergens, but the severity of asthma symptoms experienced during exposure to these insults will depend primarily on the degree of asthma control achieved in a given patient.

The long-term effects of the frequency and/or severity of asthma exacerbations are not well characterized, but it is possible that, as in COPD, asthmatics who experience severe exacerbations may show a decline in lung function. This is an area in which there are few studies and where further research is required.

Epidemiology of COPD Exacerbations

COPD will be the third leading cause of death worldwide by 2020 [33]. Many of the deaths will occur during an acute exacerbation of COPD. The true prevalence of symptomatic COPD exacerbations is very difficult to estimate, as many are not formally reported to health care providers. However, the prevalence of more severe COPD exacerbations can be estimated by visits to health practitioners or hospitalizations for episodes managed as COPD exacerbations. Thus, COPD exacerbations were responsible for $>0.9\%$ of all hospital admissions and 2.4% of the acute medical admissions in England during 2003/2004 [34]. The equivalent figures for asthma were 0.56 and 1.3%, respectively. Also, the number of COPD hospital admissions has risen considerably since the early 1990s. For example in the United Kingdom, between 1998 and 2003, the number of COPD admissions rose by 13.1% and the proportion of men admitted rose from 72 to 79%, with a mean length of stay of about 10 days [34].

Another estimate of the overall prevalence of COPD exacerbations is the frequency of visits to family physicians. In a study from Spain [35], the median number of exacerbations seen in primary care was 2/year, with 31% of patients suffering 3 or more exacerbations per year. Also, evidence collected from the General Practice Research Database in the United Kingdom suggested that COPD patients averaged 4 visits/year to the family physician in the 45–54 age range, 4.5 visits/year in the 55–64 age range, 3.6/year in the 65–74 age range, and 1.5 visits/year in the 75–84 age range because of exacerbations [36].

There is a marked season variability to COPD exacerbations, with about 50% of these occurring during the winter (fig. 1) [34]. This may be explained by the fact that most COPD exacerbations are caused by viruses (influenza virus being most commonly implicated), bacteria, or environmental pollutants and by the cold weather [37].

Conclusions

Both asthma and COPD exacerbations are major causes of morbidity and mortality in these chronic airway diseases. There are some similarities in their onset, both usually developing over 7–10 days before being recognized as an exacerbation; however, asthma exacerbations usually respond more quickly and completely to the initiation of treatment. Similarities also exist with regard to their initiating factors, particularly respiratory tract infections with viruses, although the importance of the virus types differs, with rhinovirus infection being particularly important in causing asthma exacerbations and influenza virus in COPD (although these distinctions are by no means absolute). Also, environmental allergen exposure has an important role in asthma, but not COPD. The fact that the type of virus and allergen exposure are important differences in the pathogenesis of asthma and COPD exacerbations likely accounts for the fact that the season peaks of exacerbations for asthma and COPD differ (fig. 1).

Finally, interesting differences exist in the prevalence of exacerbations in these two diseases, with 2–3 exacerbations/year being commonplace in moderate-to-severe COPD despite being on optimal treatment, but asthma exacerbations being much less frequent on optimal treatment. This is likely due to the fact that inhaled corticosteroids, the mainstay of treatment of asthma, are so effective in reducing the risk of asthma exacerbations, but have a much less obvious effect in reducing COPD exacerbations.

References

1   Fabbri L, Pauwels RA, Hurd SS: Global Strategy for the Diagnosis, Management, and Prevention of Chronic Obstructive Pulmonary Disease: GOLD Executive Summary updated 2003. COPD 2004;1:105–141.
2   Global Strategy for Asthma Management and Prevention. NIH Publication No 02-3659. Bethesda, National Institutes of Health, 2004.
3   Lane S, Molina J, Plusa T: An international observational prospective study to determine the cost of asthma exacerbations (COAX). Respir Med 2006;100:434–450.
4   Pauwels RA, Lofdahl C-G, Postma DS, O'Byrne PM, Barnes PJ, Ullman A: Effect of inhaled formoterol and budesonide on exacerbations of asthma. N Engl J Med 1997;337:1405–1411.
5   Bradley RD, Spencer GT, Semple SJ: Tracheostomy and artificial ventilation in the treatment of acute exacerbations of chronic lung disease: a study in twenty nine patients. Lancet 1964;42: 854–859.
6   O'Byrne PM, Bisgaard H, Godard PP, Pistolesi M, Palmqvist M, Zhu YJ, Ekstrom T, Bateman ED: Budesonide/formoterol combination therapy as both maintenance and reliever medication in asthma. Am J Respir Crit Care Med 2005;171:129–136.
7   Gibson PG, Wlodarczyk J, Hensley MJ, Murree-Allen K, Olson LG, Saltos N: Using quality-control analysis of peak expiratory flow recordings to guide therapy for asthma. Ann Intern Med 1995;123:488–492.
8   Rodriguez-Roisin R: Toward a consensus definition for COPD exacerbations. Chest 2000;117: 398S–401S.
9   Burge S, Wedzicha JA: COPD exacerbations: definitions and classifications. Eur Respir J Suppl 2003;41:46s–53s.
10   Tattersfield AE, Postma DS, Barnes PJ, Svensson K, Bauer CA, O'Byrne PM, Lofdahl CG, Pauwels RA, Ullman A: Exacerbations of asthma: a descriptive study of 425 severe exacerbations. The FACET International Study Group. Am J Respir Crit Care Med 1999;160:594–599.

11   Plaza V, Serrano J, Picado C, Sanchis J: Frequency and clinical characteristics of rapid-onset fatal and near-fatal asthma. Eur Respir J 2002;19:846–852.

12   Krishnan V, Diette GB, Rand CS, Bilderback AL, Merriman B, Hansel NN, Krishnan JA: Mortality in patients hospitalized for asthma exacerbations in the United States. Am J Respir Crit Care Med 2006;174:633–638.

13   Anthonisen NR, Manfreda J, Warren CP, Hershfield ES, Harding GK, Nelson NA: Antibiotic therapy in exacerbations of chronic obstructive pulmonary disease. Ann Intern Med 1987;106: 196–204.

14   Seemungal TA, Donaldson GC, Bhowmik A, Jeffries DJ, Wedzicha JA: Time course and recovery of exacerbations in patients with chronic obstructive pulmonary disease. Am J Respir Crit Care Med 2000;161:1608–1613.

15   Seemungal TA, Donaldson GC, Paul EA, Bestall JC, Jeffries DJ, Wedzicha JA: Effect of exacerbation on quality of life in patients with chronic obstructive pulmonary disease. Am J Respir Crit Care Med 1998;157:1418–1422.

16   Roberts CM, Lowe D, Bucknall CE, Ryland I, Kelly Y, Pearson MG: Clinical audit indicators of outcome following admission to hospital with acute exacerbation of chronic obstructive pulmonary disease. Thorax 2002;57:137–141.

17   Soler-Cataluna JJ, Martinez-Garcia MA, Roman SP, Salcedo E, Navarro M, Ochando R: Severe acute exacerbations and mortality in patients with chronic obstructive pulmonary disease. Thorax 2005;60:925–931.

18   Kanner RE, Anthonisen NR, Connett JE: Lower respiratory illnesses promote FEV(1) decline in current smokers but not ex-smokers with mild chronic obstructive pulmonary disease: results from the lung health study. Am J Respir Crit Care Med 2001;164:358–364.

19   Donaldson GC, Seemungal TA, Bhowmik A, Wedzicha JA: Relationship between exacerbation frequency and lung function decline in chronic obstructive pulmonary disease. Thorax 2002;57: 847–852.

20   Dowson LJ, Guest PJ, Stockley RA: Longitudinal changes in physiological, radiological, and health status measurements in alpha(1)-antitrypsin deficiency and factors associated with decline. Am J Respir Crit Care Med 2001;164:1805–1809.

21   Patil SP, Krishnan JA, Lechtzin N, Diette GB: In-hospital mortality following acute exacerbations of chronic obstructive pulmonary disease. Arch Intern Med 2003;163:1180–1186.

22   Roberts CM, Barnes S, Lowe D, Pearson MG: Evidence for a link between mortality in acute COPD and hospital type and resources. Thorax 2003;58:947–949.

23   Connors AF Jr, Dawson NV, Thomas C, Harrell FE Jr, Desbiens N, Fulkerson WJ, Kussin P, Bellamy P, Goldman L, Knaus WA: Outcomes following acute exacerbation of severe chronic obstructive lung disease. The SUPPORT investigators (Study to Understand Prognoses and Preferences for Outcomes and Risks of Treatments). Am J Respir Crit Care Med 1996;154: 959–967.

24   Groenewegen KH, Schols AM, Wouters EF: Mortality and mortality-related factors after hospitalization for acute exacerbation of COPD. Chest 2003;124:459–467.

25   Panickar JR, Dodd SR, Smyth RL, Couriel JM: Trends in deaths from respiratory illness in children in England and Wales from 1968 to 2000. Thorax 2005;60:1035–1038.

26   Getahun D, Demissie K, Rhoads GG: Recent trends in asthma hospitalization and mortality in the United States. J Asthma 2005;42:373–378.

27   Weiss KB: Seasonal trends in US asthma hospitalizations and mortality. JAMA 1990;263: 2323–2328.

28   Johnston NW, Sears MR: Asthma exacerbations. 1. Epidemiology. Thorax 2006;61:722–728.

29   Boulet LP, Cartier A, Thomson NC, Roberts RS, Dolovich J, Hargreave FE: Asthma and increases in nonallergic bronchial responsiveness from seasonal pollen exposure. J Allergy Clin Immunol 1983;71:399–406.

30   Atkinson RW, Strachan DP, Anderson HR, Hajat S, Emberlin J: Temporal associations between daily counts of fungal spores and asthma exacerbations. Occup Environ Med 2006;63:580–590.

31   Johnston SL, Pattemore PK, Sanderson G, Smith S, Lampe F, Josephs LXSP, O'Toole S, Myint SH, Tyrrell DA, et al: Community study of role of viral infections in exacerbations of asthma in 9–11 year old children. Br Med J 1995;310:1225–1229.

32 Johnston NW, Johnston SL, Duncan JM, Greene JM, Kebadze T, Keith PK, Roy M, Waserman S, Sears MR: The September epidemic of asthma exacerbations in children: a search for etiology. J Allergy Clin Immunol 2005;115:132–138.

33 Murray CJ, Lopez AD: Alternative projections of mortality and disability by cause 1990–2020: Global Burden of Disease Study. Lancet 1997;349:1498–1504.

34 Donaldson GC, Wedzicha JA: COPD exacerbations. 1. Epidemiology. Thorax 2006;61:164–168.

35 Miravitlles M, Mayordomo C, Artes M, Sanchez-Agudo L, Nicolau F, Segu JL: Treatment of chronic obstructive pulmonary disease and its exacerbations in general practice. EOLO Group. Estudio Observacional de la Limitacion Obstructiva al Flujo aEreo. Respir Med 1999;93: 173–179.

36 McGuire A, Irwin DE, Fenn P, Gray A, Anderson P, Lovering A, MacGowan A: The excess cost of acute exacerbations of chronic bronchitis in patients aged 45 and older in England and Wales. Value Health 2001;4:370–375.

37 Donaldson GC, Seemungal T, Jeffries DJ, Wedzicha JA: Effect of temperature on lung function and symptoms in chronic obstructive pulmonary disease. Eur Respir J 1999;13:844–849.

Paul M. O'Byrne, MB, FRCPI, FRCP(C), FRCPE
Firestone Institute for Respiratory Health, St. Joseph's Hospital
50 Charlton Ave East
Hamilton, ON L8N 4A6 (Canada)
Tel. +1 905 521 115, ext. 3694, Fax +1 905 521 6125, E-Mail obyrnep@mcmaster.ca

Sjöbring U, Taylor JD (eds): Models of Exacerbations in Asthma and COPD.
Contrib Microbiol. Basel, Karger, 2007, vol 14, pp 12–20

Human Rhinovirus Models in Asthma

Anne Marie Singh, William W. Busse

Department of Medicine, University of Wisconsin Medical School,
Madison, Wisc., USA

Abstract

In both children and adults, rhinovirus (RV) infections remain a major cause of exacerbations in asthma. With the use of both in vitro models of RV infection and experimental models of asthma exacerbation in humans, insight into the precise role of RV in this process has been obtained. RV infects the lower airways, and the virus itself, together with the immune response to the virus, leads to increased airway obstruction in some patients with asthma. Defects in the immune response to RV in these patients may also lead to increased symptom severity and to more significant exacerbations. Work further investigating the mechanisms of exacerbation caused by RV infection will ultimately lead to new modalities of treatment and possibly prevention of this common and significant cause of acute asthma.

Introduction

Asthma is a heterogeneous disease that ultimately leads to a clinical constellation of symptoms including cough, wheeze, and shortness of breath. These symptoms are usually accompanied by an influx of inflammatory cells and airflow obstruction. When exacerbations of asthma occur, these underlying clinical and pathophysiological features are accentuated, and a deterioration of asthma control follows, often despite existing treatment. Given the importance of asthma exacerbations to a loss of disease control, and patient risks for increased morbidity and mortality, it is important to understand both the immunopathogenesis of these events and mechanisms leading to increases in asthma severity.

Many factors contribute to an increase in asthma symptoms including respiratory infections, allergens, irritants, and occupational exposures. Each of these exacerbating factors likely acts through different mechanisms, but has a final common pathway that includes cellular inflammation, enhanced bronchial responsiveness, and greater airflow obstruction [1]. As respiratory viruses are the most frequent cause of both the common cold and asthma exacerbations [2, 3], this area of study holds important clinical insights. In addition to causing the exacerbation itself, viral infections may also alter cytokine expression even when the asthmatic airway is already inflamed [4]. This response may lead to decreased viral clearance and greater persistence of inflammation [5]. Thus, the study of viral infections in asthma will provide further insight into both immunopathogenesis and direction of future treatment.

Epidemiology

Respiratory viruses are the most frequent cause of both the common cold and asthma exacerbations. In a study of children aged 9–11 years, 80–85% of asthma exacerbations that resulted in reduced peak expiratory flow and wheezing were found due to viral upper respiratory infections [2]. Similarly, Nicholson et al. [3] found that 57% of adult asthma exacerbations were due to upper respiratory tract infections. Of the infections associated with virus-induced exacerbations, rhinovirus (RV) was the most commonly found microorganism (~65% of all cases). Also, admission rates for asthma exacerbations correlate significantly with the seasonal pattern of RV infections, e.g., September through December and April [6]. Thus, RV is the major cause of asthma exacerbations and respiratory infection-provoked asthma attacks in both children and adults.

Microbiology

RV, a member of the Picornaviridae family, is a small, single-stranded RNA virus with a capsid of four distinct proteins. There are more than 100 serotypes of RVs, the majority of which bind to the intercellular adhesion molecule-1 (ICAM-1) receptor. Epithelial cells are the principal target of RV, the site of viral replication, and the initiators of the immune/inflammatory response [7]. When RV infects epithelial cells, cytokines and other mediators are generated and released (see below) to recruit cells to the airway. In the acute phase of a cold, neutrophils are the primary inflammatory infiltrate found. Plasma leakage also occurs to contribute to airway edema and a build-up of airway secretions. Although not fully established, airway obstruction with virus-provoked asthma likely arises from inflammatory cell infiltration, edema, secretions, and mucus hypersecretion. Other factors, including direct neuronal stimulation by the virus, may also promote bronchospasm and airflow obstruction [8].

As RV infections are the major cause of the common cold, it was initially felt that those infections were confined to the upper airway. Now, however, it is clear that RV can, and does, infect the lower airway as well [9]. When wild-type RV isolates were incubated at 33°C (upper airway temperature) versus 37°C (lower airway temperature), viral replication was observed at both temperatures [10]. Also, RV appeared to infect both upper and lower segments of the respiratory tree equally well [11] and the frequency of lower airway infection was similar to that observed in the upper airway [12]. In addition, analysis of sputum samples after experimental infection with strain RV-16 have indicated that the amount of virus in the lower airway varies among individuals, but can reach the same high levels of infection found in upper airway tissues [11]. These findings suggest RV can indeed infect the lower airway. Differences in the host response to the virus may account for the variability in lower airway viral infection and, by extension, chest symptoms in different patients. It is proposed that worsening of asthma symptoms arises secondarily to the extension of infection to the lower airway where inflammation increases and airflow obstruction follows.

Experimental Virus Infection in Human Asthmatic Patients

Experimental virus infection of humans may be a useful tool to understand basic mechanisms in viral infections and how they may relate to asthma exacerbations. Patients infected with an experimental safety-tested strain of RV often have a relatively mild response compared to those with naturally occurring infections, including less severe clinical and cellular responses [5]. In addition, by infecting patients in a clinical study, early, prospective analysis of both symptoms and physiological responses can be accomplished in a monitored setting.

Immunopathogenesis
Patients with asthma or allergic rhinitis suffer from increased pathophysiology as a result of RV infection. Although cold symptoms usually last 1 week or less, decreases in peak flow following RV infections can persist for a median of 2 weeks in school-aged children [2]. This association suggests that RV infection persists in the lower airway in asthmatic patients or that the consequences of an RV infection (such as lower airway features of asthma) last beyond the initial phase of the infection [13].

As the RV infection becomes established in the airway epithelium, the virus-infected tissue and airway leukocytes release a variety of cytokines and mediators that are likely to cause or enhance airway inflammation. It is probable that the generation of mediators, rather than a direct airway injury by the RV itself, enhances existing airway inflammation. Increased transcription of the

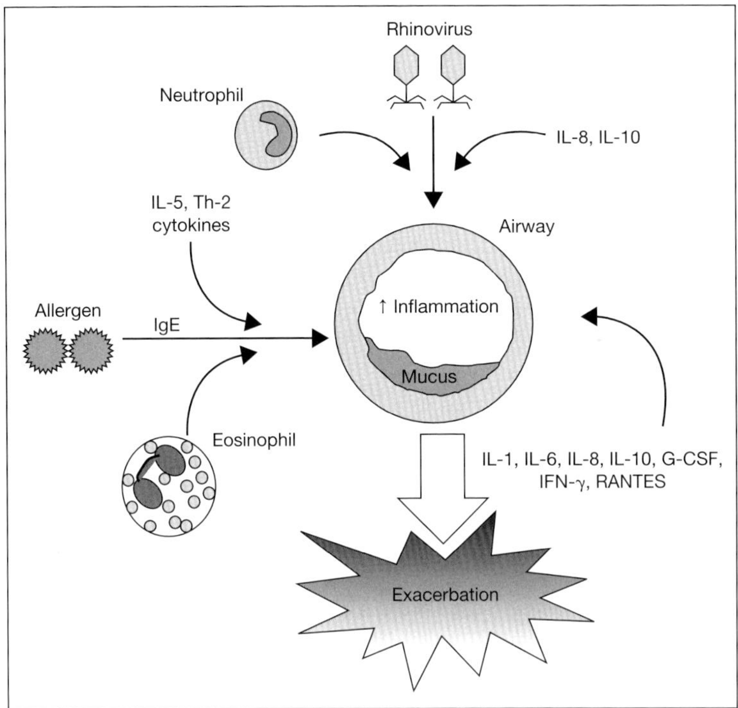

Fig. 1. Infection with RV causes neutrophil influx with increased inflammatory cytokines. Once activated, epithelial cells release further mediators (IL-1, IL-6, IL-8, IL-10, G-CSF, IFN-γ, RANTES) which increase inflammation, and lead to exacerbation. This response is increased in the presence of allergic airway inflammation.

genes encoding many cytokines and chemokines occur with RV infection, including IL-1, IL-6, IL-8, IL-10, IL-16, G-CSF, interferon-γ (IFN-γ) and RANTES [9, 14], which have been described both in vitro in studies of bronchial epithelial cell lines and in vivo (fig. 1). In addition, Papadopoulus et al. [15] found an increase in the Th2 cytokines IL-4 and IL-13 in asthmatic subjects. Induction of inflammatory cell recruitment and activation may occur as a result of these mediators. RV infections, like other viral infections, cause a transient increase in the number of circulating neutrophils, a response that corresponds to symptoms of the upper respiratory infection [12]. These cytokines can also increase synthesis of leukocytes to further enhance leukocyte recruitment to the airway, and, perhaps, activate neutrophils to cause inflammation. Also, viremia has been found early during an acute asthma exacerbation caused by RV, suggesting a causal role in the pathogenesis of the exacerbation [16].

The role of individual cytokines or chemokines in the immunopathogenesis is not clear and the subject of ongoing investigation. IL-1, TNF-α, and IL-6 induce the acute phase of the response and play a role in the activation of both T and B lymphocytes. Production of IL-8 by epithelial cells results in neutrophil recruitment, a hallmark of viral induced asthma exacerbations. Increases in sputum neutrophils correlate with increases in sputum IL-8 [17]. In addition, increases in IL-8 have been found in children with natural colds [18] and experimental RV-16 infections of asthmatic subjects [19, 20]. GMCSF, eotaxin and RANTES play a role in eosinophil survival and chemotaxis [21].

The role of IL-10 in virus-induced exacerbations has shown contrasting results, and deserves special consideration. Grissell et al. [22] completed a study of 59 patients over the age of 7 years who were admitted to the hospital with acute asthma. The group was divided into those with and without evidence of a viral infection. Patients with viral infection (RV was again the predominant infection) and an asthma exacerbation had significant increases in IL-10 and RANTES. The authors proposed that the generation of these factors might explain some of the characteristics of inflammation in virus-induced asthma. For example, IL-10 can suppress eosinophil cellular infiltration, whereas RANTES promotes eosinophil degranulation and neutrophil influx. After recovery from the viral illness and asthma exacerbation, IL-10 levels returned to baseline values, and sputum eosinophilia reappeared. These observations led the authors to suggest that virus-induced asthma has immune response features distinct from allergen-provoked asthma, and that IL-10 likely has a unique but undefined role in this process [22].

In contrast, Corne et al. [14] measured cytokine, chemokine and mediator levels in nasal lavage from 44 adults (23 of whom were atopic) taken during acute and convalescent phases of a common cold illness. In the acute phase of illness, IL-10 levels were significantly higher in the nonatopic group compared to the atopic group. There was no significant difference in IL-10 levels during the convalescent phase. As IL-10 has been shown to inhibit some proinflammatory cytokines, it was hypothesized that increased IL-10 during the acute phase of infection downregulated and limited the virus-induced inflammatory response [14]. According to these authors, prolonged inflammatory responses occur as a direct result of this failure to mount an appropriate IL-10 response in atopic subjects.

Asthma is also associated with an increase in ICAM-1 expression [23]. ICAM-1 is a principal receptor for the majority of RVs, and thus, upregulation and activation of ICAM-1, which occur both with chronic allergen exposure and with RV infection, may explain an increased susceptibility of asthma patients to RV infection.

In addition, recent work has demonstrated that the innate immune response to infection with RV may be restricted and deficient in some asthmatic patients [24]. For example, Wark et al. [24] found that bronchial epithelial cells obtained from asthmatic patients and then infected with RV-16 were resistant to early apoptosis and had a profoundly and significantly deficient type I interferon response compared to RV-infected cells from nonasthmatic patients. Although there was no difference in expression of ICAM-1, IL-6, or RANTES in this setting, there was increased viral RNA expression and late cell lysis in the airway tissues from asthmatic patients. The RV-infected cells from asthmatic patients lived longer and demonstrated decreased apoptosis. Because of these immune defects in host defense, asthmatic patients may be more likely to develop a persistent lower airway infection.

As a result of these findings, the authors investigated whether interferon-β (IFN-β) may be an underlying mechanism that contributes to an abnormal antiviral response in asthma. Interestingly, data confirmed that IFN-β induction by RV infection of bronchial epithelial cells from asthmatic subjects was profoundly impaired and related to decreased apoptosis of epithelial cells and increased viral replication [24]. Thus, asthmatic patients may have decreased innate immunity to RV that results in increased cell lysis (causing a more defective epithelial layer and inflammation) and viral replication, but decreased infected cell apoptosis. A defect in IFN-β may be responsible for these effects.

In another study, 22 subjects with asthma or allergic rhinitis were inoculated with RV-16, and proliferation and cytokine production from peripheral blood mononuclear cells were determined before, 7 and 28 days after inoculation. Parry et al. [25] and Gern et al. [20] found that peak RV shedding and symptom severity were inversely proportional to IFN-γ generation and directly proportional to IL-5 production by peripheral blood mononuclear cells. Thus, asthma patients with a more robust Th1 response were more effcctive at clearing the illness, and vice versa. Current data suggest that some asthmatic subjects likely have dysfunctional antiviral interferon responses to RV, and these abnormalities may make these patients more susceptible to an infection and the consequences of the illness (fig. 2).

Effect of RV and Allergic Sensitization
RV infection may also act synergistically with allergic inflammation. For example, Calhoun et al. [13] demonstrated that patients with allergic rhinitis, who were infected with experimental RV-16, had enhanced histamine release into bronchial alveolar lavage fluid immediately following antigen challenge, and this response was followed by increased eosinophil recruitment 48 h later. The increase in airway eosinophils in response to antigen challenge persisted for up to 1 month postinfection. In addition, Grünberg et al. [19] found that RV-infected patients had increased airway responsiveness to an inhaled histamine challenge. Conversely, when atopic patients were challenged with nasal

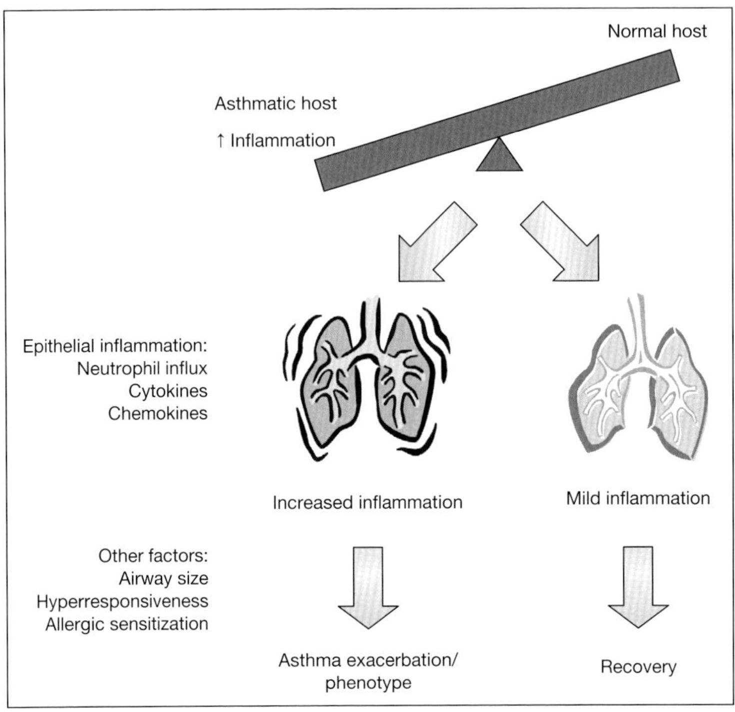

Fig. 2. The antiviral response of IFN-γ and lymphocyte infiltration results in more rigorous epithelial inflammation in the asthmatic host. An increase in cytokines/chemokines and neutrophil influx in combination with other factors gives rise to exacerbation.

allergens prior to inoculation of RV, the onset of symptoms was delayed with less severe responses [19]. In addition, Murray et al. [26] found that the combination of virus detection and sensitization with high allergen exposure in children substantially increased the risk of hospitalization compared with virus detection or sensitization and allergen exposure alone (OR 19.4, 3.7–101.5; $p < 0.001$). In a case control study of 60 adult hospitalized patients, it was found that being sensitized and exposed to allergens was an independent risk factor for being admitted to the hospital (OR 2.3) [27]. However, the combination of sensitization, high exposure and viral detection considerably increased the risk of being admitted to the hospital with asthma (OR 8.4, 2.1–32.8) [27]. Therefore, it appears that RV and allergen-specific responses act together. This relationship may be stronger when RV infection occurs prior to allergen challenge.

Conclusion

RV is a frequent and important contributor to exacerbations of asthma. It has been through studies of experimental models of infection that the role of RV in asthma exacerbation has been clarified. By its immune response to virus and generation of mediators, it has become clear that the epithelial cell has a prominent role in the immune response, which then furthers inflammation, leading to the constellation of chest symptoms and airway obstruction. Ongoing studies will continue to further the understanding of the immunopathogenesis of viral-induced asthma, and will help contribute to future treatments of the most common cause of exacerbations.

References

1 Busse WW, O'Bryne PM, Holgate ST: Asthma pathogenesis; in Adkinson NF Jr, Yunginger JW, Busse WW, Bochner BS, Holgate ST, Simons FER (eds): Middleton's Allergy Principles and Practice, ed 6. St Louis, Mosby, 2003, pp 1175–1207.

2 Johnston SL, Pattemore PK, Sanderson G, Smith S, Lampe F, Josephs L, Symington P, O'Toole S, Myint SH, Tyrrell DA, Holgate ST: Community study of role of viral infections in exacerbations of asthma in 9–11 year old children. BMJ 1995;310:1225–1229.

3 Nicholson KG, Kent J, Ireland DC: Respiratory viruses and exacerbations of asthma in adults. BMJ 1993;307:982–986.

4 Brooks GD, Buchta KA, Swenson CA, Gern JE, Busse WW: Rhinovirus-induced interferon-gamma and airway responsiveness in asthma. Am J Respir Crit Care Med 2003;168:1091–1094.

5 Message SD, Johnston SL: The immunology of virus infection in asthma. Eur Respir J 2001;18:1013–1025.

6 Johnston SL, Pattemore PK, Sanderson G, Smith S, Campbell MJ, Josephs LK, Cunningham A, Robinson BS, Myint SH, Ward ME, Tyrrell DA, Holgate ST: The relationship between upper respiratory infections and hospital admissions for asthma: a time-trend analysis. Am J Respir Crit Care Med 1996;154:654–660.

7 Papadopoulos NG, Papi A, Psarras S, Johnston SL: Mechanisms of rhinovirus-induced asthma. Paediatr Respir Rev 2004;5:255–260.

8 Bowerfind WM, Fryer AD, Jacoby DB: Double-stranded RNA causes airway hyperreactivity and neuronal M2 muscarinic receptor dysfunction. J Appl Physiol 2002;92:1417–1422.

9 Friedlander SL, Busse WW: The role of rhinovirus in asthma exacerbations. J Allergy Clin Immunol 2005;116:267–273.

10 Papadopoulos NG, Sanderson G, Hunter J, Johnston SL: Rhinoviruses replicate effectively at lower airway temperatures. J Med Virol 1999;58:100–104.

11 Mosser AG, Brockman-Schneider R, Amineva S, Burchell L, Sedgwick JB, Busse WW, Gern JE: Similar frequency of rhinovirus-infectible cells in upper and lower airway epithelium. J Infect Dis 2002;185:734–743.

12 Gern JE: Rhinovirus respiratory infections and asthma. Am J Med 2002;112(suppl 6A):19S–27S.

13 Calhoun WJ, Dick EC, Schwartz LB, Busse WW: A common cold virus, rhinovirus 16, potentiates airway inflammation after segmental antigen bronchoprovocation in allergic subjects. J Clin Invest 1994;94:2200–2208.

14 Corne JM, Lau L, Scott SJ, Davies R, Johnston SL, Howarth PH: The relationship between atopic status and IL-10 nasal lavage levels in the acute and persistent inflammatory response to upper respiratory tract infection. Am J Respir Crit Care Med 2001;163:1101–1107.

15 Papadopoulos NG, Stanciu LA, Papi A, Holgate ST, Johnston SL: A defective type 1 response to rhinovirus in atopic asthma. Thorax 2002;57:328–332.
16 Xatzipsalti M, Kyrana S, Tsolia M, Psarras S, Bossios A, Laza-Stanca V, Johnston SL, Papadopoulos NG: Rhinovirus viremia in children with respiratory infections. Am J Respir Crit Care Med 2005;172:1037–1040.
17 Pizzichini MM, Pizzichini E, Efthimiadis A, Chauhan AJ, Johnston SL, Hussack P, Mahony J, Dolovich J, Hargreave FE: Asthma and natural colds. Inflammatory indices in induced sputum: a feasibility study. Am J Respir Crit Care Med 1998;158:1178–1184.
18 Teran LM, Johnston SL, Schroder JM, Church MK, Holgate ST: Role of nasal interleukin-8 in neutrophil recruitment and activation in children with virus-induced asthma. Am J Respir Crit Care Med 1997;155:1362–1366.
19 Grünberg K, Timmers MC, Smits HH, de Klerk EP, Dick EC, Spaan WJ, Hiemstra PS, Sterk PJ: Effect of experimental rhinovirus 16 colds on airway hyperresponsiveness to histamine and interleukin-8 in nasal lavage in asthmatic subjects in vivo. Clin Exp Allergy 1997;27:36–45.
20 Gern JE, Vrtis R, Grindle KA, Swenson C, Busse WW: Relationship of upper and lower airway cytokines to outcome of experimental rhinovirus infection. Am J Respir Crit Care Med 2000;162:2226–2231.
21 Gleich GJ: Mechanisms of eosinophil-associated inflammation. J Allergy Clin Immunol 2000; 105:651–663.
22 Grissell TV, Powell H, Shafren DR, Boyle MJ, Hensley MJ, Jones PD, Whitehead BF, Gibson PG: Interleukin-10 gene expression in acute virus-induced asthma. Am J Respir Crit Care Med 2005;172:433–439.
23 Vignola AM, Campbell AM, Chanez P, Bousquet J, Paul-Lacoste P, Michel FB, Godard P: HLA-DR and ICAM-1 expression on bronchial epithelial cells in asthma and chronic bronchitis. Am Rev Respir Dis 1993;148:689–694.
24 Wark PA, Johnston SL, Bucchieri F, Powell R, Puddicombe S, Laza-Stanca V, Holgate ST, Davies DE: Asthmatic bronchial epithelial cells have a deficient innate immune response to infection with rhinovirus. J Exp Med 2005;201:937–947.
25 Parry DE, Busse WW, Sukow KA, Dick CR, Swenson C, Gern JE: Rhinovirus-induced PBMC responses and outcome of experimental infection in allergic subjects. J Allergy Clin Immunol 2000;105:692–698.
26 Murray CS, Poletti G, Kebadze T, Morris J, Woodcock A, Johnston SL, Custovic A: A study of modifiable risk factors for asthma exacerbations: virus infection and allergen exposure increase the risk of asthma hospitalization in children. Thorax 2006;61:376–382.
27 Green RM, Custovic A, Sanderson G, Hunter J, Johnston SL, Woodcock A: Synergism between allergens and viruses and risk of hospital admission with asthma: case-control study. BMJ 2002;324:763.

William W. Busse, MD
Department of Medicine, Section of Allergy
Pulmonary and Critical Care Medicine
University of Wisconsin Medical School
Madison, WI 53792 (USA)
Tel. +1 608 263 6183, Fax +1 608 263 3104, E-Mail wwb@medicine.wisc.edu

Sjöbring U, Taylor JD (eds): Models of Exacerbations in Asthma and COPD.
Contrib Microbiol. Basel, Karger, 2007, vol 14, pp 21–32

Allergen Inhalation Challenge: A Human Model of Asthma Exacerbation

Gail M. Gauvreau, Michelle Y. Evans

Department of Medicine, McMaster University, Hamilton, Ont., Canada

Abstract

Allergen challenge by inhalation is a very useful clinical and research tool for evaluating allergic airway disease. Inhalation of allergen leads to cross-linking of allergen-specific IgE bound to IgE receptors on mast cells and basophils. This is followed by activation of secretory pathways to release preformed and newly generated mediators of bronchoconstriction and vascular permeability. The onset of bronchoconstriction, representing the early phase of the asthmatic response, can be detected within 10 min of the inhalation, reaches a maximum within 30 min, and resolves within 3 h. The late-phase asthmatic response starts between 4 and 8 h, and is characterized by cellular inflammation of the airway, increased bronchiovascular permeability, and mucus secretion. The late-phase asthmatic response is also associated with increased airway responsiveness to nonallergic stimuli. Approximately half of the allergic asthmatic patients develop a late-phase response after allergen inhalation challenge. There has been a tremendous interest in trying to understand the differences between the pathways leading to the dual response and those leading to the early response alone. The current hypotheses are discussed in this chapter. Our understanding of the allergen inhalation challenge model and the complex interplay between leukocytes, tissue and inflammatory mediators will doubtlessly help to define novel and relevant targets for new drugs for the treatment of allergic asthma.

Methodology

Allergen challenge is a very useful clinical and research tool for evaluating allergic disease. Allergen can be administered by the inhaled, nasal or cutaneous route, providing a means to confirm an allergic status, investigate the mechanisms of allergic disease, or evaluate the efficacy of allergy therapy. This chapter will describe the methodology of the allergen inhalation challenge (AIC) model in subjects with allergic asthma, summarize the complex cellular mechanisms involved in the development of allergen-induced airway responses, and review the effects of commonly used asthma therapies on these airway events.

Although it is usually well tolerated, controlled AIC may induce dangerous side effects and must therefore be carried out with caution. AIC can induce anaphylaxis and severe acute bronchoconstriction, as well as exacerbation of asthma, with recurrent nocturnal symptoms lasting for several days. The degree of bronchial hyperresponsiveness to nonallergic stimuli such as histamine and methacholine, and the circulating levels of specific IgE levels are the main determinants of early-phase bronchial responsiveness to allergen [1, 2]. Thus, the allergen-induced early asthmatic response can be accurately predicted using the histamine PC_{20} [defined as the concentration causing a 20% fall of the forced expiratory volume in 1 s (FEV_1)], and a skin test endpoint [2]. A recent review of the relationship between allergen PC_{20}, the methacholine PC_{20}, and the skin test endpoint showed that histamine and methacholine-based formulae both predicted allergen PC_{20} within 3 doubling concentrations in over 92% of the subjects studied [3].

Methods of allergen inhalation have not been uniform across research laboratories. One commonly used method uses tidal breathing of doubling concentrations of allergen from a Wright nebulizer [4]. The starting concentration is chosen from the results of a methacholine challenge test and skin prick tests of increasing concentrations of allergen extracts; the lowest concentration of allergen causing a 2-mm skin wheal and the methacholine PC_{20} is determined to calculate the starting concentration of inhaled allergen. Other laboratories deliver cumulative allergen inhalations from a dosimeter [5]. Although single allergen dose challenges can also achieve the desired responses [6], safety concerns support a step-wise approach where the response to allergen can be monitored between increasing doses.

With any of these methodologies used to deliver allergen, AIC induces bronchoconstriction with acute onset within 10 min of the inhalation, reaches a maximum within 30 min, and resolves back to baseline within 3 h. This bronchoconstrictor response is called the early-phase asthmatic response, and is considered to be positive if the FEV_1 falls by 15–20% from preinhalation baseline. Some of the patients who develop an early-phase asthmatic response to AIC also develop a late-phase asthmatic response, which can be measured by a fall in FEV_1 of at least 15% from preinhalation baseline, starting 4–8 h after challenge. The prevalence of the late-phase asthmatic response in adults who had developed an early-phase response after inhalation of ragweed pollen extract is approximately 50% [7, 8], and this holds true for other allergen extracts including house dust and cat dander. Notably, the late-phase asthmatic response causes a more prominent and sustained increase in bronchospasm than the early-phase asthmatic response, and is associated with airway hyperresponsiveness to nonallergic stimuli such as histamine and methacholine.

Local airway challenge has also been safely performed in subjects with asthma by direct instillation of challenge solution to the selected segmental airways using a bronchoscope. Subjects are typically pretreated with nebulized salbutamol, atropine and midazolam. In addition, oxygen is delivered via nasal cannulae throughout the procedure and oxygen saturation and heart rate are monitored for safety purposes. The allergen extract used for segmental allergen challenge is that causing the largest wheal response on skin prick testing, and the concentration used for the endobronchial challenge is one tenth of that which elicited a skin wheal with a diameter of 3 mm during a skin wheal dose-response series. This method is particularly useful for assessing inflammation in restricted areas of the airway, but cannot be used to assess the early asthmatic response and late asthmatic response accurately due to the invasive nature of the procedure.

Mechanisms of the Early-Phase Asthmatic Response

Initiation of the acute allergen-induced bronchoconstriction phase requires circulating IgE to inhaled allergens, such as dust mite or pollen proteins. Inhalation of allergen leads to interaction of the allergenic epitopes with mast cell or basophil-bound allergen-specific IgE, resulting in cross-linking of IgE receptors. The signals induced by cross-linking result in activation of secretory pathways to release both preformed and newly generated mediators of bronchoconstriction and vascular permeability, including histamine, leukotrienes C4/D4, prostaglandin D_2, and platelet-activating factor, as well as of a wide array of cytokines and chemokines (fig. 1).

The cysteinyl leukotrienes (CysLTs), LTC_4, LTD_4 and LTE_4, are potent mediators of bronchoconstriction [9], and are released by resident mast cell and basophils into the airways after AIC [10]. Urinary LTE_4 excretion is a good marker of the rate of total body production of CysLTs, and therefore elevated urinary LTE_4 levels are found in subjects who develop an early asthmatic response [11]. CysLTs are believed responsible for most of the bronchoconstriction measured during the early asthmatic response. This has been confirmed experimentally by blocking the response with CysLTs receptor antagonists (LTRA) [12] or CysLTs synthesis inhibitors [13]. Histamine also contributes to bronchoconstriction during the early-phase asthmatic response, as shown by superior protection or by combination of LTRA and antihistamine treatment [14]. Left untreated, the early-phase asthmatic response resolves within 2 h of onset.

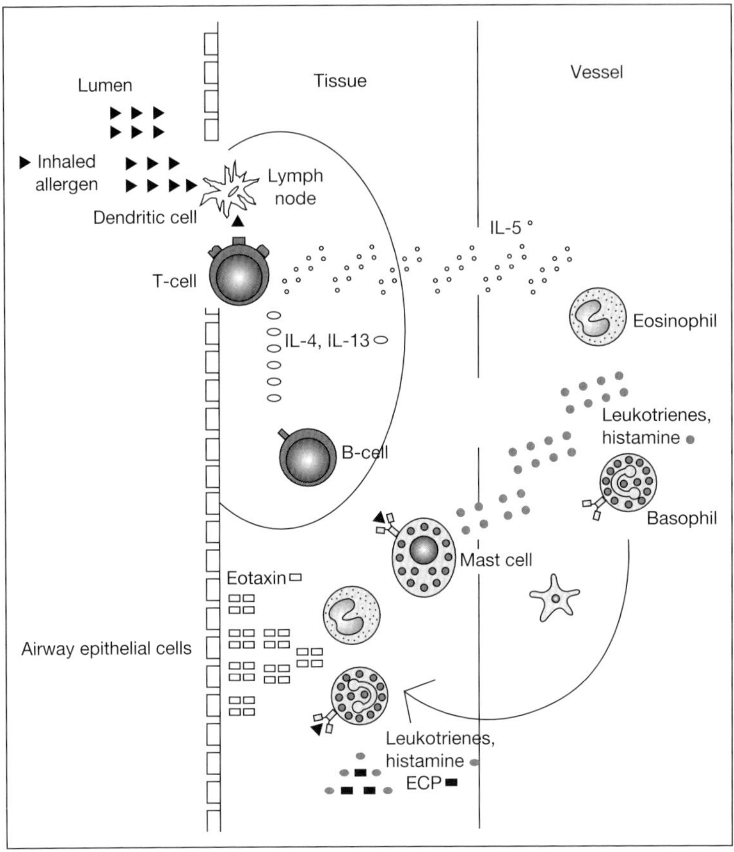

Fig. 1. Interactions between leukocytes, tissue and cell mediators. Allergen cross-links IgE receptors on mast cells and basophils, inducing release of pre-formed and newly synthesized mediators. Dendritic cells capture antigen and present to T lymphocytes, which clonally expand and release Th2 cytokines. Eosinophils and basophils are activated by cytokines such as IL-5 and attracted by chemokines such as eotaxin. ECP = Eosinophilic cationic protein.

Mechanisms of the Late-Phase Asthmatic Response

During the late-phase asthmatic response, mediators thought to originate from mast cells can also be measured at levels similar to those observed during the early-phase response [11, 15]. This suggests that a second phase of mediator release occurs during the late-phase response, which may drive the development of a second response. However, unlike the early-phase asthmatic response, the effects of mediator release are not likely to act upon smooth muscle alone.

This is because the late-phase asthmatic response is very prolonged and is not readily reversed by inhaled bronchodilators acting on smooth muscle β-adrenergic receptors (β-agonists) or muscarinic receptors (anticholinergics). The idea therefore developed that the late-phase asthmatic response is caused by an inflammatory response in the airways, and that it is associated with other causes of reduced airflow, including mucosal edema and secretions. The second phase of mediator release from IgE-bearing cells leads to attraction of inflammatory cells into the airways.

Morphologically, the late-phase asthmatic response is characterized by cellular inflammation of the airway, increased bronchovascular permeability, and increased mucus secretion. The airway cellular infiltrate various leukocytes, including eosinophils, mast cells, basophils, and T lymphocytes [16–19]. The airway inflammation associated with allergic asthma is initiated through a complex interaction of antigen-presenting cells and T lymphocytes resulting in the release of a cascade of cytokines and chemokines regulating the progress of the allergic inflammatory response (fig. 1).

Eosinophils are believed to be critical proinflammatory cells in airway mucosal damage in asthma. AIC in atopic asthmatic subjects is associated with recruitment and activation of eosinophils in the airways. Once activated, eosinophils release toxic products, including eosinophil cationic protein, able to damage bronchial structures and to increase bronchial hyper-responsiveness.

The correlation between eosinophil numbers and CysLT concentrations after allergen challenge is consistent with these cells being a principal source of CysLTs within the airways during and after the late-phase asthmatic response. Eosinophil activation, as assessed by secretion of eosinophil cationic protein, and granulocyte/macrophage colony-stimulating factor (GM-CSF) levels are significantly increased in bronchoalveolar lavage fluid and cells, consistent with the hypothesis that eosinophils, regulated by GM-CSF, contribute to allergen-induced decreases in airway function [20].

Allergen-induced increases in sputum eosinophils are associated with the presence of cytokines and chemokines specifically involved in activation and chemotaxis of eosinophils [17]. Eotaxin is a C-C chemokine with selective activity for eosinophils and basophils. Allergen-induced increases in eotaxin-positive cells are correlated with increases in eosinophils suggesting that eotaxin may contribute to allergen-induced recruitment of eosinophils to the airway in asthmatic subjects [21]. Increases in activated T lymphocytes, eosinophils, and elevation of the mRNA expression of interleukin-5 (IL-5) and GM-CSF in bronchial biopsies after AIC in atopic asthmatics support the view that cytokines/chemokines possibly released by activated T lymphocytes may contribute to local eosinophil accumulation during allergen-induced asthmatic responses [22].

Mast cells and basophils have also been implicated in the late-phase asthmatic response. The ratio of degranulating to granulated mast cells in airway biopsy tissue is higher in subjects with a late-phase response as compared to subjects without a late-phase response, and the number of mast cells significantly correlates with the severity of the late-phase response [23]. Similarly, the number of sputum basophils measured 24 h after AIC is weakly correlated with the late-phase asthmatic response [18]. This suggests that secretion of pro-inflammatory mediators including histamine and CysLTs by mast cells and basophils, as well as secretion of large amounts of IL-4 and IL-13 by basophils following IgE-mediated activation, may contribute to the observed physiological changes in the airways.

T lymphocytes have been identified as a key factor for the development of the late-phase asthmatic response. After AIC, T lymphocytes have been shown to be clonally expanded in the lower respiratory tract, demonstrating the inhalation of sensitizing allergens can recruit allergen-specific T cell clones to the lung [24]. Furthermore, cytokines produced by activated Th2-type CD4+ T cells in the airway, such as IL-4 and IL-5, may contribute to late asthmatic responses by mechanisms that include eosinophil accumulation [25].

Following AIC, the development of late-phase asthmatic responses typically occur following IgE-mediated early-phase asthmatic responses. However, late-phase asthmatic responses can also be generated independent of IgE following intradermal or inhaled administration of allergen-derived T cell peptide epitopes. Late-phase asthmatic response occurring without a proceeding early-phase asthmatic response supports the view that inhalation of allergen-derived T cell peptide epitopes can elicit an isolated late-phase response in sensitized subjects by an IgE-independent mechanism. The resulting airway inflammation measured is similar to that following inhalation of allergen extracts, and suggests that peptide- and whole allergen-induced late-phase asthmatic responses share a common mechanism associated with increases in CysLTs and the infiltration of inflammatory cells [26].

Nonspecific Airway Hyperresponsiveness

The increase in nonspecific airway hyperresponsiveness that is associated with the late-phase asthmatic response has been of particular interest in the literature. This was investigated in the context of AIC, following Dr. Altounyan's observation that patients with asthma showed increases in airway responsiveness also to stimuli other than the allergen itself, at the times of natural allergen exposure during pollen season. It was found that large changes in airway responsiveness can occur for a prolonged period of time in patients with asthma who underwent

AIC [27]. Further investigations have shown that bronchial responsiveness to inhaled histamine and methacholine can be observed as early as several hours after AIC. Although the etiology of this phenomenon is unclear, increased neural activity was thought to be involved since allergen also induces increases in airway responsiveness to the mast cell stimulus adenosine $5'$-monophosphate. However, since there is no change in responsiveness to sodium metabisulphite, an indirect neural stimulus with similar characteristics to bradykinin [28], activation of airway sensory nerves is unlikely to contribute to the increase in airway responsiveness following inhalation of allergen.

Several cell types that accumulate in the allergen-challenged airway may play a role in the development of airway hyperresponsiveness. A weak positive correlation is demonstrated between airway eosinophil [29] and basophil [18] percentages, and increased nonallergic bronchial reactivity observed after allergen challenge. This suggests that accumulation of allergen-induced migratory cells may be partially associated with the degree of airway hyperresponsiveness.

Isolated Early Responders and Dual Responders

Several independent laboratories have reported that approximately half of the allergic asthmatic patients develop a late-phase response after AIC [4, 7, 8, 30]. There has been a tremendous interest in trying to understand the differences between the isolated early responders and dual responders. Separating the mechanisms of individual responses could provide useful information for pharmacological management of these patients. To date, several differences have been reported between the two phenotypes.

Examination of the airways has confirmed that those who develop a late-phase asthmatic response also have higher levels of inflammatory infiltrate, as well as a more pronounced degree of airway hyperresponsiveness [18]. The development of a late-phase asthmatic response is associated with an early recruitment of eosinophils and epithelial desquamation in the airways. The different time course of allergen-induced airway inflammation between allergen-induced early and dual responders may also reflect the induction of CD4+ T lymphocytes expressing IL-10 in early but not dual responders [31]. In addition, it has been noted that after AIC, CD4+ and CD8+ IFN-γ-positive cells in peripheral blood significantly decrease in dual responders, whereas these cell types, along with IL-12-positive cells significantly increase in the airways of isolated early responders [32]. Therefore differences in activation of T lymphocytes could be crucial for the development of isolated early responses and dual responses.

Another aspect of allergic inflammatory responses is the induction of inflammatory progenitor cells, which contribute to disease through the continued

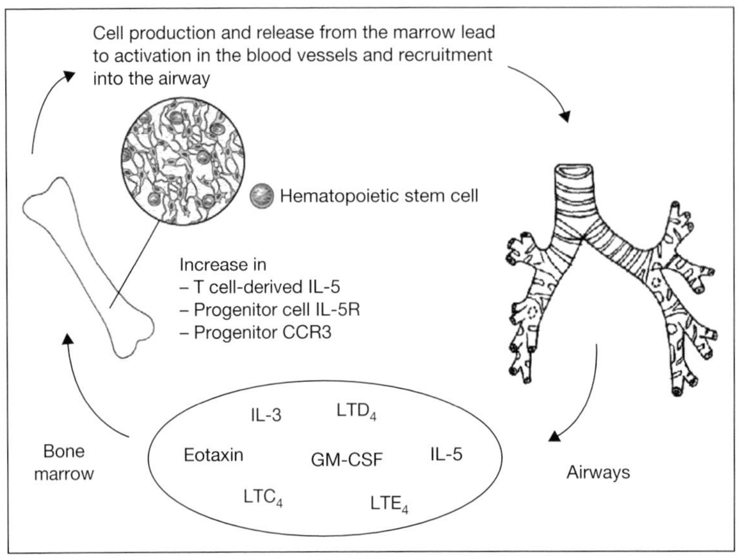

Fig. 2. Allergen-induced signals from the airway stimulate the release of eosinophil/basophil progenitor cells, and induce the expression of receptors for eotaxin and IL-5. T lymphocytes migrate to the bone marrow and produce increased amounts of IL-5.

production of inflammatory effector cells. In dual responders, there is a significant increase in the numbers of eosinophil/basophil colony-forming units in the bone marrow 24 h after an allergen challenge as well as increased responsiveness of these cells to IL-5, a cytokine specifically involved in eosinophil growth and maturation. This is not observed in those with only an isolated early response after allergen inhalation [33, 34]. Furthermore, dual responders have more IL-3-responsive progenitors detected as early as 5 h after allergen inhalation, and more IL-5-responsive progenitor cells were detected at 12 and 24 h after AIC. Again, these changes are not detected in isolated early responders. This suggests that there is allergen-induced activation of an eosinophilopoietic process that is rapid and sustained in dual responders. Inhaled allergen also causes trafficking of T lymphocytes to the bone marrow in subjects who develop late-phase asthmatic responses, and there is a significant increase in the ability of bone marrow cells, particularly T lymphocytes, to produce IL-5 [35]. The relationship between increased bone marrow IL-5 levels and increased eosinophil production reflects time-dependent changes in cytokine levels in the bone marrow, which controls the differentiation of eosinophil/basophil progenitors (fig. 2).

Other investigators have focused on the concentrations of allergen extracts which elicit a response. Subjects in whom the early reaction was induced by a

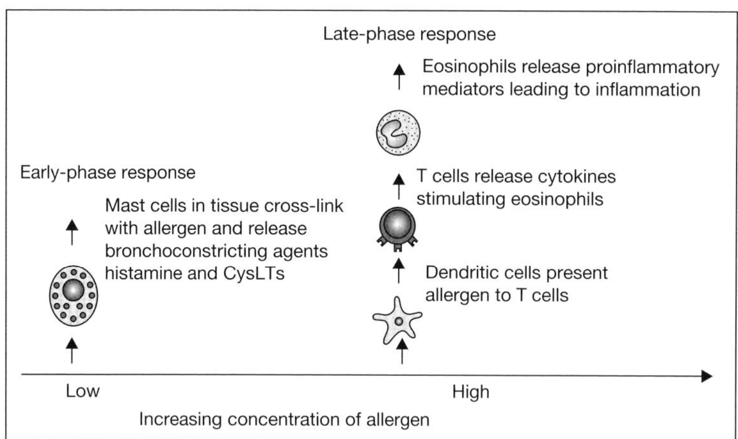

Fig. 3. Increasing concentrations of allergen inhaled into the airway may be required for the generation of the late-phase asthmatic response.

low dose of inhaled antigen were found to be more likely to develop a late response, and it can be concluded that the occurrence of a late-phase asthmatic response probably depends on the dose of allergen administered but is independent of the degree of nonspecific bronchial responsiveness to methacholine prior to the challenge [30, 36]. Furthermore, when subjects are pretreated with antihistamine to allow a greater dose of inhaled allergen, it is possible to induce a late asthmatic response in subjects who previously demonstrated only an early response [37]. A scenario follows whereby exceeding a threshold dose of allergen activates other cell types, probably dendritic cells and T lymphocytes, thereby driving the late-phase asthmatic response (fig. 3).

Allergen Inhalation Challenge for Testing Drug Efficacy

AIC are most commonly used as an investigative tool to study the pathophysiology of asthma and possible blocking effects of immunotherapy. When a new pharmaceutical agent is studied, its ability to block the early- and late-phase asthmatic responses and subsequent hyperresponsiveness is presumptive evidence for effectiveness. Within this study design, the same individuals are challenged with exactly the same dose of allergen after receiving treatment with the test drug, and after receiving treatment with placebo. The endpoint measurements in such studies are the maximum early and late percent decreases in the FEV_1, and the areas under the curve for the early response (0–2 h postchallenge) and the late response (3–7 h postchallenge). The reproducibility of measurements is such that fewer than 8 subjects are required to show 50% attenuation of either the early or late response, with a 90% power [38]. The reproducibility of allergen-induced sputum eosinophils has also been investigated. With a randomized crossover study design, the sample size predicted to be necessary to observe 50% attenuation of allergen-induced percent of eosinophils with a power of 0.95 was less than 10 subjects [39]. In terms of the late-phase asthmatic response, the AIC is also able to differentiate a single dose of an active inhaled corticosteroid from placebo and a highly potent inhaled corticosteroid from a weak inhaled corticosteroid [40], making this a very robust model for testing asthma therapies.

Conclusions

Since the early studies with careful documentation of the early- and late-phase asthmatic responses to AIC, much has been learned about the clinical significance of exposure to a sensitizing allergen. The inhaled allergen challenge model will likely continue to be extensively employed for studying the pathogenesis of airway inflammation, and for assessment of new therapeutics in early clinical proof-of-concept studies. Our evolving understanding of the AIC model and the complex interplay between leukocytes, tissue and inflammatory mediators will doubtlessly help to define novel and relevant targets for the treatment of allergic asthma.

References

1   van der Veen MJ, Lopuhaa CE, Aalberse RC, Jansen HM, van der Zee JS: Bronchial allergen challenge with isolated major allergens of *Dermatophagoides pteronyssinus*: the role of patient characteristics in the early asthmatic response. J Allergy Clin Immunol 1998;102:24–31.
2   Cockcroft DW, Murdock KY, Kirby J, Hargreave F: Prediction of airway responsiveness to allergen from skin sensitivity to allergen and airway responsiveness to histamine. Am Rev Respir Dis 1987;135:264–267.
3   Cockcroft DW, Davis BE, Boulet LP, Deschesnes F, Gauvreau GM, O'Byrne PM, Watson RM: The links between allergen skin test sensitivity, airway responsiveness and airway response to allergen. Allergy 2005;60:56–59.
4   O'Byrne PM, Dolovich J, Hargreave FE: Late asthmatic responses. Am Rev Respir Dis 1987;136:740–751.
5   Frolund L, Svendsen UG, Nielsen NH, Weeke B, Madsen F: Bronchial allergen challenge: comparison between two different methods of provocation. Clin Allergy 1987;17:439–448.
6   Tamura G, Mue S, Ishihara T, Takishima T: The single exposure method for inhalation challenge with allergen. J Allergy Clin Immunol 1985;75:47–54.
7   Robertson DG, Kerigan AT, Hargreave FE, Chalmers R, Dolovich J: Late asthmatic responses induced by ragweed pollen allergen. J Allergy Clin Immunol 1974;54:244–254.

8   Booij-Noord H, de Vries K, Sluiter HJ, Orie NG: Late bronchial obstructive reaction to experimental inhalation of house dust extract. Clin Allergy 1972;2:43–61.

9   Dahlen SE, Hedqvist P, Hammarstrom S, Samuelsson B: Leukotrienes are potent constrictors of human bronchi. Nature 1980;288:484–486.

10   Wenzel SE, Westcott JY, Larsen GL: Bronchoalveolar lavage fluid mediator levels 5 minutes after allergen challenge in atopic subjects with asthma: relationship to the development of late asthmatic responses. J Allergy Clin Immunol 1991;87:540–548.

11   Manning PJ, Rokach J, Malo JL, Ethier D, Cartier A, Girard Y, Charleson S, O'Byrne PM: Urinary leukotriene E4 levels during early and late asthmatic responses. J Allergy Clin Immunol 1990;86: 211–220.

12   Hamilton A, Faiferman I, Stober P, Watson RM, O'Byrne PM: Pranlukast, a cysteinyl leukotriene receptor antagonist, attenuates allergen-induced early- and late-phase bronchoconstriction and airway hyperresponsiveness in asthmatic subjects. J Allergy Clin Immunol 1998;102: 177–183.

13   Hamilton AL, Watson RM, Wyile G, O'Byrne PM: Attenuation of early and late phase allergen-induced bronchoconstriction in asthmatic subjects by a 5-lipoxygenase activating protein antagonist, BAYx 1005. Thorax 1997;52:348–354.

14   Davis BE, Todd DC, Cockcroft DW: Effect of combined montelukast and desloratadine on the early asthmatic response to inhaled allergen. J Allergy Clin Immunol 2005;116:768–772.

15   O'Sullivan S, Roquet A, Dahlen B, Dahlen S, Kumlin M: Urinary excretion of inflammatory mediators during allergen-induced early and late phase asthmatic reactions. Clin Exp Allergy 1998;28:1332–1339.

16   Rossi GA, Crimi E, Lantero S, Gianiorio P, Oddera S, Crimi P, Brusasco V: Late-phase asthmatic reaction to inhaled allergen is associated with early recruitment of eosinophils in the airways. Am Rev Respir Dis 1991;144:379–383.

17   Gauvreau GM, Watson RM, O'Byrne PM: Kinetics of allergen-induced airway eosinophilic cytokine production and airway inflammation. Am J Respir Crit Care Med 1999;160:640–647.

18   Gauvreau GM, Lee JM, Watson RM, Irani AM, Schwartz LB, O'Byrne PM: Increased numbers of both airway basophils and mast cells in sputum after allergen inhalation challenge of atopic asthmatics. Am J Respir Crit Care Med 2000;161:1473–1478.

19   Burastero SE, Crimi E, Balbo A, Vavassori M, Borgonovo B, Gaffi D, Frittoli E, Casorati G, Rossi GA: Oligoclonality of lung T lymphocytes following exposure to allergen in asthma. J Immunol 1995; 155:5836–5846.

20   Woolley KL, Adelroth E, Woolley MJ, Ellis R, Jordana M, O'Byrne PM: Effects of allergen challenge on eosinophils, eosinophil cationic protein, and granulocyte-macrophage colony-stimulating factor in mild asthma. Am J Respir Crit Care Med 1995;151:1915–1924.

21   Zeibecoglou K, Macfarlane AJ, Ying S, Meng Q, Pavord I, Barnes NC, Robinson DS, Kay AB: Increases in eotaxin-positive cells in induced sputum from atopic asthmatic subjects after inhalational allergen challenge. Allergy 1999;54:730–735.

22   Bentley AM, Meng Q, Robinson DS, Hamid Q, Kay AB, Durham SR: Increases in activated T lymphocytes, eosinophils, and cytokine mRNA expression for interleukin-5 and granulocyte/macrophage colony-stimulating factor in bronchial biopsies after allergen inhalation challenge in atopic asthmatics. Am J Respir Cell Mol Biol 1993;8:35–42.

23   Crimi E, Chiaramondia M, Milanese M, Rossi GA, Brusasco V: Increased numbers of mast cells in bronchial mucosa after the late-phase asthmatic response to allergen. Am Rev Respir Dis 1991;144:1282–1286.

24   Borgonovo B, Casorati G, Frittoli E, Gaffi D, Crimi E, Burastero SE: Recruitment of circulating allergen-specific T lymphocytes to the lung on allergen challenge in asthma. J Allergy Clin Immunol 1997;100:669–678.

25   Robinson D, Hamid Q, Bentley A, Ying S, Kay AB, Durham SR: Activation of CD4+ T cells, increased TH2-type cytokine mRNA expression, and eosinophil recruitment in bronchoalveolar lavage after allergen inhalation challenge in patients with atopic asthma. J Allergy Clin Immunol 1993;92:313–324.

26   Ali FR, Oldfield WL, Higashi N, Larche M, Kay AB: Late asthmatic reactions induced by inhalation of allergen-derived T cell peptides. Am J Respir Crit Care Med 2004;169:20–26.

27 Cockcroft DW, Ruffin RE, Dolovich J, Hargreave FE: Allergen-induced increase in non-allergic bronchial reactivity. Clin Allergy 1977;7:503–513.
28 Evans DJ, Coulby LJ, O'Connor BJ: Effect of allergen challenge on airway responsiveness to histamine and sodium metabisulphite in mild asthma. Thorax 1996;51:1185–1191.
29 Oddera S, Silvestri M, Penna R, Galeazzi G, Crimi E, Rossi GA: Airway eosinophilic inflammation and bronchial hyperresponsiveness after allergen inhalation challenge in asthma. Lung 1998;176:237–247.
30 MacIntyre D, Boyd G: Factors influencing the occurrence of a late reaction to allergen challenge in atopic asthmatics. Clin Allergy 1984;14:311–317.
31 Matsumoto K, Gauvreau GM, Rerecich T, Watson RM, Wood LJ, O'Byrne PM: IL-10 production in circulating T cells differs between allergen-induced isolated early and dual asthmatic responders. J Allergy Clin Immunol 2002;109:281–286.
32 Yoshida M, Watson RM, Rerecich T, O'Byrne PM: Different profiles of T-cell IFN-gamma and IL-12 in allergen-induced early and dual responders with asthma. J Allergy Clin Immunol 2005;115:1004–1009.
33 Sehmi R, Wood LJ, Watson R, Foley R, Hamid Q, O'Byrne PM, Denburg JA: Allergen-induced increases in IL-5 receptor alpha-subunit expression on bone marrow-derived CD34+ cells from asthmatic subjects. A novel marker of progenitor cell commitment towards eosinophilic differentiation. J Clin Invest 1997;100:2466–2475.
34 Dorman SC, Sehmi R, Gauvreau GM, Watson RM, Foley R, Jones GL, Denburg JA, Inman MD, O'Byrne PM: Kinetics of bone marrow eosinophilopoiesis and associated cytokines after allergen inhalation. Am J Respir Crit Care Med 2004;169:565–572.
35 Wood LJ, Sehmi R, Dorman S, Hamid Q, Tulic MK, Watson RM, Foley R, Wasi P, Denburg JA, Gauvreau G, O'Byrne PM: Allergen-induced increases in bone marrow T lymphocytes and interleukin-5 expression in subjects with asthma. Am J Respir Crit Care Med 2002;166:883–889.
36 Machado L, Stalenheim G: Factors influencing the occurrence of late bronchial reactions after allergen challenge. Allergy 1990;45:268–274.
37 Lai CK, Beasley R, Holgate ST: The effect of an increase in inhaled allergen dose after terfenadine on the occurrence and magnitude of the late asthmatic response. Clin Exp Allergy 1989;19:209–216.
38 Inman MD, Watson R, Cockcroft DW, Wong BJ, Hargreave FE, O'Byrne PM: Reproducibility of allergen-induced early and late asthmatic responses. J Allergy Clin Immunol 1995;95:1191–1195.
39 Gauvreau GM, Watson RM, Rerecich TJ, Baswick E, Inman MD, O'Byrne PM: Repeatability of allergen-induced airway inflammation. J Allergy Clin Immunol 1999;104:66–71.
40 Inman MD, Watson RM, Rerecich T, Gauvreau GM, Lutsky BN, Stryszak P, O'Byrne PM: Dose-dependent effects of inhaled mometasone furoate on airway function and inflammation after allergen inhalation challenge. Am J Respir Crit Care Med 2001;164:569–574.

Dr. Gail Gauvreau
Health Sciences Center, Room 3U25, McMaster University
1200 Main St West
Hamilton, ON L8N 3Z5 (Canada)
Tel. +1 905 525 9140, ext. 22791, Fax +1 905 528 1807, E-Mail gauvreau@mcmaster.ca

Sjöbring U, Taylor JD (eds): Models of Exacerbations in Asthma and COPD.
Contrib Microbiol. Basel, Karger, 2007, vol 14, pp 33–41

Cellular and Animals Models for Rhinovirus Infection in Asthma

Maria Xatzipsalti, Nikolaos G. Papadopoulos

Allergy Department, 2nd Pediatric Clinic, University of Athens, Athens, Greece

Abstract

Human rhinoviruses (RVs) are responsible for the majority of upper respiratory tract infections. Despite the high prevalence, the pathogenesis is incompletely understood. Experimental models would permit study of the immunological response to infections. Animal models have many limitations because of the anatomic and physiological differences between mammalian species. The only nonhuman animals susceptible to RV are chimpanzees and gibbons. Mouse models are not used because of host cell tropism of RV. This problem may have been partially overcome by transfecting mouse cells with viral RNA, by replacing mouse ICAM-1 with the human counterpart and by using a variant virus. It remains to be seen if these advances will translate into establishment of useful mouse models. In the absence of animal models, epithelial cell lines such as BEAS-2B, A549, 16HBE and HEp-2 have been used. Fibroblasts and smooth muscle cells have also been used. Although transformed cell lines have many properties in common with normal epithelial cells, they lose certain differentiated functions. Therefore, primary and recently well-differentiated cultures are used to study the immune response. In addition to a local inflammatory response, a systemic immune response to RV does develop; therefore peripheral blood mononuclear cells and dendritic cells have been infected with RV, shedding additional light on cell-mediated immunity. Cellular models are invaluable investigational tools for understanding mechanisms of RV-induced asthma and evaluating new targets for therapy.

Copyright © 2007 S. Karger AG, Basel

Introduction

Human rhinoviruses (RVs) are responsible for the majority of upper respiratory tract infections (common cold). These infections are generally mild in healthy populations but pose a serious health risk for patients with chronic respiratory diseases. The common cold is implicated in approximately half of the asthma attacks in adults [1] and in 80–85% of asthma exacerbations in children [2]. Therefore, the overall morbidity and economic burden attributable to RV is considerable. Despite the high prevalence of RV infections, the pathogenesis of the subsequent disease process is incompletely understood. Several lines of evidence indicate that production of mediators, recruitment of cells and changes in the biochemistry of cells during infection are of importance, and therefore the understanding of the events involved in the pathogenesis of RV infections is crucial for the development of improved therapeutic approaches.

The design of therapeutically useful antiviral drugs is currently hampered by our limited knowledge of the cellular and molecular mechanisms of RV infections. Valid experimental models would permit studies of the immunological response to infections and perhaps determine the key mediators of the disease, identifying new therapeutic targets. A model of asthma exacerbation in humans has already been developed on the basis of experimental RV infection, as described in a separate chapter. Although ideal, in that its results are more or less directly applicable to human disease, the human model is hampered by its inherent practical difficulty as well as limitations in the extent of possible interventions. On the other hand, animal models have a number of limitations because of their anatomic and physiological differences in the respiratory tract of different mammalian species. Up to now, the development of cellular models has therefore been of considerable importance in increasing our understanding of the disease.

Animal Models

The only nonhuman animals susceptible to RV and that have been considered as models are the chimpanzee and the gibbon. In 1972, gibbons and chimpanzees were experimentally infected with RV in order to test the antiviral effect of orally administered drugs (triazinoindole and bis-benzimidazoles). RV was isolated in throat swabs from infected animals and serum antiviral titers were measured postinfection for correlation with the virus isolation data. Nevertheless, these animal models are impractical, as they are costly and difficult to work with and they were therefore not used for many years. However, recently the chimpanzee model was used to demonstrate the in vivo efficacy of tICAM453, a truncated, soluble form of intercellular adhesion molecule-1 (ICAM-1). tICAM453 was known to be a potent inhibitor of RV in vitro and using this animal model the investigators demonstrated that the intranasal application of tICAM453 was efficient in preventing RV infection in chimpanzees [3].

Attempts to design murine models of RV infection were never abandoned. The major obstacle to the development of mouse models is the host cell tropism of RV. Approximately 10% of RV serotypes can use both the human and mouse forms of low-density lipoprotein receptor, and a mouse model capitalizing on this fact was established [4]. However, the remaining 90% of RV use human

ICAM-1 to attach and enter the cells. These viruses do not bind to mouse ICAM-1 and therefore infection of mice or of mouse bronchial epithelial cells with the major RV variants is not possible. This problem can be overcome by transfecting mouse cells with viral RNA [5]. Mouse 1 and 2 domains of ICAM-1 (chimerical ICAM-1), known to be involved in RV binding, were replaced with the equivalent human domains. When cells expressing this chimerical construct were inoculated with RV16, infection and virus replication occurred, as evidenced by the appearance of cytopathic effect following infection, and increasing virus titers. Using this cellular model it was also demonstrated that respiratory epithelial cells of both human and mouse species have similar capacity to support replication of RV. It remains to be established whether humanized transgenic mice expressing these chimerical receptors on bronchial epithelial cells will allow RV infection similar to that seen in humans.

Other investigators have found that virus replication may be dependent on a species-specific cellular mechanism. Thus a variant virus, 16/L, was isolated after alternate passage of RV16 between human HeLa cells and ICAM-1 expressing mouse L cells and was demonstrated to produce higher levels of infection in mouse cells [5, 6]. The variant obtained in this manner had three amino acid changes identified in the protein 2C, a protein that influences membrane vesicle trafficking, membrane permeability and viral RNA replication. Another report using the same passage techniques described that simultaneous changes in the 2B and 3A proteins, involved in RV replication, conferred the ability of a serotype 39 strain to replicate efficiently in mouse cells [7]. Identification of RV isolates displaying a broader host range may facilitate the investigation of cellular proteins required for efficient viral growth and the development of a murine model of RV infections [5].

Recently, low but existent viral replication as well as inflammatory events similar to those seen in humans were observed in a murine model [Johnston, pers. commun.]. The detailed results from this model are awaited.

In vitro Models

Respiratory epithelial cells are the primary sites of RV infections in humans. In the absence of appropriate animal models of RV infection, investigators have used epithelial cell lines to study the mechanisms of RV-induced exacerbations of asthma. A transformed continuous cell line, BEAS-2B cells, derived from normal human bronchial epithelial cells after transfection with an adenovirus 12-SV40 hybrid virus, was among the first in which RV infection could be established [8]. Cells grown close to confluence are exposed to RV at appropriate concentrations. They are then incubated for some time, with constant motion (rotating wall vessel culture or gentle shaking) at room temperature. Virus particles are detected by exposing monolayers of HeLa cells in 96-well plates to serial 10-fold dilutions of virus-containing medium. The viral shedding into the culture medium is assessed by the cytopathic effects of this medium on HeLa cells. With the use of this model it was demonstrated that RV is able to induce cytokine/chemokine secretion, including IL-8, IL-6 and GM-CSF through NF-κB activation. These cytokines/chemokines have biological properties that are of interest with respect of the pathogenesis of colds. IL-8 is a potent chemoattractant and activator of neutrophils, IL-6 can induce B cell differentiation and T cell activation and GM-CSF can prime both neutrophils and eosinophils for enhanced activation to chemical stimuli [9]. In addition RV can induce the expression of ICAM-1, which is not only the site of attachment for the 90% of RVs but also plays a vital role in the recruitment and migration of immune effector cells at the site of infection through the interaction with the two β_2 integrins, leukocyte function-associated antigen (LFA-1; CD11a/CD18) and Mac-1 (CD11b/CD18). Pretreatment of cells with Th1 and Th2 cytokines affects the expression of ICAM-1; this may account for the different frequency of upper respiratory infections in healthy and asthmatic subjects. RV infection of bronchial epithelium results also in inducing eosinophil recruitment [10], probably through the production of the chemokines RANTES, MIP-1α, eotaxin, eotaxin-2, MCP-2, MCP-3 and MCP-4. Activation of p38 kinase was shown recently to be a key event in the regulation of the virus-induced cytokine transcription [11]. VCAM-1, an endothelial adhesion protein with a central role in recruitment and activation of lymphocytes and eosinophils through interactions with their β_1 and/or β_7 integrins, is also induced by RV infection via NF-κB- and GATA-mediated transcriptional upregulation [12].

Investigators have also used several other epithelial cell lines for studying RV infections. A549 cells, derived from a human alveolar cell carcinoma with properties of type II alveolar epithelial cells [13], 16HBE, an SV40 large T antigen-transformed human airway epithelial cell line [14], and HEp-2 cells [15] are widely used. In accordance with the findings in BEAS-2B cells, RV infection causes a significant increase in cytokines and chemokines in these cell lines.

Although the bronchial epithelium forms a barrier between the airway lumen and the underlying cells and is actively involved in the immune response to RV through the production of a variety of immunomodulatory factors, nonepithelial cells of mesenchymal origin may also play a major role. The presence of RV has been detected in subepithelial cells by in situ hybridization, suggesting that the virus could spread to fibroblasts via infected epithelial cells and therefore these cells may participate in the inflammatory response. Human fetal lung fibroblasts (MRC-5 strain) were infected with RV, and the production of inflammatory mediators was assessed. In these cells, RV was found to stimulate the production of the cytokines IL-6, IL-11, and VEGF and the chemokines IL-8, ENA-78 and RANTES [16], but not eotaxin [17]. Whether the different cytokine secretion patterns displayed by different cell types translate to a differential potential for cell recruitment in vivo is not known.

Human airway smooth muscle cells were also infected with RV. Previous experiments were carried out using rabbit airway smooth muscle cells and cell lines but very recently primary muscle cells (human cultured airway smooth muscle cells, HASM) were isolated from bronchial tissues or lung tissues and cultured according to standard methods. These cells were then infected with RV and the proinflammatory cytokine response was measured. No differences were seen in RV replication between cells from asthmatic patients and nonasthmatic subjects, in contrast to the findings in normal bronchial epithelial cells (see below). Furthermore, in contrast to epithelial cells RV-induced IL-6 was greater in the asthmatic-derived HASM cells in comparison to cells derived from nonasthmatic subjects. IL-8 secretion was induced only in asthmatics while eotaxin was not found to be induced either in healthy or in asthmatic cells [18, 19]. Taking into account that smooth muscle cells are further away from the epithelial surface, the major infection site, the clinical relevance of these differences is difficult to interpret; nevertheless, it seems that a differential response between cells from normal and atopic/asthmatic individuals extends to several structural cell types. Whether this is an inherent defect of these cells or a result of reprogramming through exposure to immune mediators is still unknown.

In addition to the local inflammatory response, it has been shown that a systemic immune response to RV does develop. In order to model such a response, peripheral blood mononuclear cells (PBMC) have been used. PBMC were infected with RV leading to an increase in the expression of the early activation marker CD69 on T cells. RV also induced the secretion of IFN-γ from both peripheral blood T cells and NK cells [20], as well as the production of IL-8 [21]. Exposure of PBMC from normal and atopic asthmatic subjects to RV resulted in an upregulation of IFN-γ, IL-12, and IL-10 production in both groups although the IFN-γ response was considerably lower in asthmatics [22]. Upregulation of IL-4 was only seen in cells from asthmatics. Furthermore, the immune response as measured by IFN-γ and IL-5 secretion from infected PBMC from atopic asthmatic patients before an experimentally induced cold was inversely correlated with the viral shedding after inoculation [23], suggesting that variations in mononuclear cell responses to RV could contribute to the individual variability in viral shedding during RV infections in subjects with respiratory allergy or asthma. The described abnormal immune response of atopic asthmatic individuals to RV was shown to influence the epithelial response to virus. In a more complicated model, PBMCs were exposed to RV and the resulting supernatants, containing a multitude of factors associated with the immune response to the virus, were used to treat epithelial cells before or during RV infection. When such conditioned media from normal or atopic asthmatic individuals were used, it was shown that in the 'atopic environment' epithelial inflammation was reduced, while virus replication and the resulting epithelial cytotoxicity were significantly increased (unpubl. data from our laboratory).

Antigen presentation and costimulation mediated by dendritic cells (DCs) have also been studied. Monocytes stimulated by RV were shown to inhibit the allostimulatory capacity of myeloid DCs, by producing large amounts of the immunosuppressive IL-10 [24]. Recently, it was shown that DCs infected by RV demonstrate a diminished T cell stimulatory capacity, inducing a deep anergic state through two receptors B7-H1 and sialoadhesin (Sn) [25], shedding light on novel features of cell-mediated immunity.

Although human epithelial cell lines share many properties in common with normal epithelial cells, transformed cell lines may lose certain differentiated functions, resulting in an altered response to RV infections. Therefore, investigators examined whether primary cultures of epithelial cells can also be infected with RV. Initially human tracheal epithelial cells, obtained 3–6 h after death from patients, were used. RV infection was demonstrated by an increase of the viral titer in supernatants and of viral RNA in the cell extract. RV infection increased the production of cytokines such as IL-6 and IL-8 and upregulated the expression of ICAM-1, as shown previously in cell lines. However, infection of human tracheal epithelial cells with RV also caused increases in the production of IL-1β and TNF-α, which differs from RV infection of BEAS-2B cells [8]. Human nasal epithelial cells (HNE) isolated from nasal tissues obtained from patients undergoing ethmoidectomy or surgery for polyposis or turbinate hypertrophy have also been used. Although HNE cells represent the first site of RV infections it has been shown recently that RV can replicate and infect the lower airways [26] producing lower airway inflammation by increasing the neutrophils, T cells and eosinophils in bronchial lavage fluid [10] and enhancing epithelial expression of intracellular adhesion molecules. For this reason, human bronchial epithelial cells were also cultured either from surgical specimens of human bronchi or from bronchial brushings. It was demonstrated that RV could indeed replicate in these cells in vitro by the increase in RV titer and infection by the cytopathic effect (increased granularity, cell rounding, and detachment) on epithelial cells. Primary human lower airway cells produced a similar inflammatory response after RV infection and reproduced the results of epithelial cell lines, as they were able to produce increased amounts of IL-8, IL-6, RANTES, ICAM-1, VCAM-1, VEGF [27, 28] and IP-10 [29] after RV infection.

Both primary cells and cell lines represent highly undifferentiated cells; for example they generally do not form tight junctions. Therefore results obtained in such cultures may not represent the in vivo situation. In an attempt to address this issue, cells form tracheas and nasal scrapings were grown under different culture conditions to achieve a range of differentiation from a squamous to a fully pseudostratified phenotype [30]. These cultures were then infected with RV and their susceptibility was compared. The well-differentiated cells showed increased resistance to RV; therefore higher multiplicity of infection and longer infection time were used. RV infection had little effect on the levels of expression of RANTES, IL-6, IL-8, TNF-α or eotaxin-2 and eotaxin-3, in differentiated cells in contrast to the increased expression shown in undifferentiated cultures. In addition, no change in the expression of ICAM-1 was demonstrated after RV infection in well-differentiated cells, in contrast to primary bronchial epithelial cells, probably due to the existence of tight junctions [30]. Nevertheless, the major antiviral systems, which are normally increased after RV infection, were almost similar. RV induced IFN-β secretion, in agreement with primary epithelial cells [31] through the PKR-IFN-β-JAK-STAT pathway.

Recent data from studies in cultured epithelial cells suggest that there are fundamental differences in the response to RV infection such that viral RNA expression and late virus release into supernatant were increased 50- and 7-fold, respectively, in asthmatic cells compared with healthy controls. The difference may be due to an impairment of virus-induced INF-β mRNA expression in asthmatic cultures which produced >2.5 times less INF-β protein than cells cultured from normal individuals [31]. In addition, in comparison with cells from normals both primary bronchial epithelial cells and alveolar macrophages from asthmatic individuals showed deficient induction of INF-λ by RV, and the response was highly correlated with disease severity [32].

Conclusions

As RV are the major cause of asthma exacerbations a detailed understanding of the molecular and cellular mechanisms of RV infection is required to enable the development of novel measurements for prevention and treatment. While an appropriate animal model of RV-induced asthma would greatly help in this direction, cellular models are still invaluable investigational tools for both understanding mechanisms and evaluating new targets for therapy.

References

1   Nicholson KG, Kent J, Ireland DC: Respiratory viruses and exacerbations of asthma in adults. BMJ 1993;307:982–986.
2   Johnston SL, Pattemore PK, Sanderson G, Smith S, Lampe F, Josephs L, Symington P, O'Toole S, Myint SH, Tyrrell DAJ, Holgate ST: Community study of role of viral infections in exacerbations of asthma in 9–11 year old children. BMJ 1995;310:1225–1228.

3   Huguenel ED, Cohn D, Dockum DP, Greve JM, Fournel MA, Hammond L, Irwin R, Mahoney J, McClelland A, Muchmore E, Ohlin AC, Scuderi P: Prevention of rhinovirus infection in chimpanzees by soluble intercellular adhesion molecule-1. Am J Respir Crit Care Med 1997;155: 1206–1210.

4   Yin FH, Lomax NB: Establishment of a mouse model for human rhinovirus infection. J Gen Virol 1986;67:2335–2340.

5   Harris JR, Racaniello VR: Changes in rhinovirus protein 2C allow efficient replication in mouse cells. J Virol 2003;77:4773–4780.

6   Tuthill TJ, Papadopoulos NG, Jourdan P, Challinor LJ, Sharp NA, Plumpton C, Shah K, Barnard S, Dash L, Burnet J, Killington RA, Rowlands DJ, Clarke NJ, Blair ED, Johnston SL: Mouse respiratory epithelial cells support efficient replication of human rhinovirus. J Gen Virol 2003;84: 2829–2836.

7   Harris JR, Racaniello VR: Amino acid changes in proteins 2B and 3A mediate rhinovirus type 39 growth in mouse cells. J Virol 2005;79:5363–5373.

8   Subauste MC, Jacoby DB, Richards SM, Proud D: Infection of a human respiratory epithelial cell line with rhinovirus. Induction of cytokine release and modulation of susceptibility to infection by cytokine exposure. J Clin Invest 1995;96:549–557.

9   Papadopoulos NG, Papi A, Psarras S, Johnston SL: Mechanisms of rhinovirus-induced asthma. Paediatr Respir Rev 2004;5:255–260.

10 Fraenkel DJ, Bardin PG, Sanderson G, Lampe F, Johnston SL, Holgate ST: Lower airways inflammation during rhinovirus colds in normal and in asthmatic subjects. Am J Respir Crit Care Med 1995;151:879–886.

11 Griego SD, Weston CB, Adams JL, Tal-Singer R, Dillon SB: Role of p38 mitogen-activated protein kinase in rhinovirus-induced cytokine production by bronchial epithelial cells. J Immunol 2000;165:5211–5220.

12 Papi A, Johnston SL: Respiratory epithelial cell expression of vascular cell adhesion molecule-1 and its up-regulation by rhinovirus infection via NF-kappaB and GATA transcription factors. J Biol Chem 1999;274:30041–30051.

13 Lieber M, Smith B, Szakal A, Nelson-Rees W, Todaro G: A continuous tumor-cell line from a human lung carcinoma with properties of type II alveolar epithelial cells. Int J Cancer 1976;17: 62–70.

14 Bodo M, Baroni T, Bellocchio S, Calvitti M, Lilli C, D'Alessandro A, Muzi G, Lumare A, Abbritti G: Bronchial epithelial cell matrix production in response to silica and basic fibroblast growth factor. Mol Med 2001;7:83–92.

15 Reischl A, Reithmayer M, Winsauer G, Moser R, Gosler I, Blaas D: Viral evolution toward change in receptor usage: adaptation of a major group human rhinovirus to grow in ICAM-1-negative cells. J Virol 2001;75:9312–9319.

16 Ghildyal R, Dagher H, Donninger H, de Silva D, Li X, Freezer NJ, Wilson JW, Bardin PG: Rhinovirus infects primary human airway fibroblasts and induces a neutrophil chemokine and a permeability factor. J Med Virol 2005;75:608–615.

17 Meyer JE, Stangenberg S, Weise JB, Beck C, Schmidt C, Kurz K, Beier UH, Maune S: Nasal RANTES and eotaxin production pattern in response to rhinovirus infection. Rhinology 2006;44: 140–144.

18 Oliver BG, Johnston SL, Baraket M, Burgess JK, King NJ, Roth M, Lim S, Black JL: Increased proinflammatory responses from asthmatic human airway smooth muscle cells in response to rhinovirus infection. Respir Res 2006;7:71.

19 Johnson PR, Armour CL, Carey D, Black JL: Heparin and PGE2 inhibit DNA synthesis in human airway smooth muscle cells in culture. Am J Physiol 1995;269:L514–L519.

20 Gern J, Vrtis R, Kelly E, Dick E, Busse W: Rhinovirus produces nonspecific activation of lymphocytes through a monocyte-dependent mechanism. J Immunol 1996;157:1605–1612.

21 Johnston SL, Papi A, Monick MM, Hunninghake GW: Rhinoviruses induce interleukin-8 mRNA and protein production in human monocytes. J Infect Dis 1997;175:323–329.

22 Papadopoulos NG, Stanciu LA, Papi A, Holgate ST, Johnston SL: A defective type 1 response to rhinovirus in atopic asthma. Thorax 2002;57:328–332.

23 Parry DE, Busse WW, Sukow KA, Dick CR, Swenson C, Gern JE: Rhinovirus-induced PBMC responses and outcome of experimental infection in allergic subjects. J Allergy Clin Immunol 2000;105:692–698.

24 Stockl J, Vetr H, Majdic O, Zlabinger G, Kuechler E, Knapp W: Human major group rhinoviruses downmodulate the accessory function of monocytes by inducing IL-10. J Clin Invest 1999;104: 957–965.

25 Kirchberger S, Majdic O, Steinberger P, Bluml S, Pfistershammer K, Zlabinger G, Deszcz L, Kuechler E, Knapp W, Stockl J: Human rhinoviruses inhibit the accessory function of dendritic cells by inducing sialoadhesin and B7-H1 expression. J Immunol 2005;175:1145–1152.

26 Papadopoulos NG, Bates PJ, Bardin PG, Papi A, Leir SH, Fraenkel DJ, Meyer J, Lackie PM, Sanderson G, Holgate ST, Johnston SL: Rhinoviruses infect the lower airways. J Infect Dis 2000;181:1875–1884.

27 De Silva D, Dagher H, Ghildyal R, Lindsay M, Li X, Freezer NJ, Wilson JW, Bardin PG: Vascular endothelial growth factor induction by rhinovirus infection. J Med Virol 2006;78:666–672.

28 Volonaki E, Psarras S, Xepapadaki P, Psomali D, Gourgiotis D, Papadopoulos NG: Synergistic effects of fluticasone propionate and salmeterol on inhibiting rhinovirus-induced epithelial production of remodelling-associated growth factors. Clin Exp Allergy 2006;36:1268–1273.

29 Spurrell JC, Wiehler S, Zaheer RS, Sanders SP, Proud D: Human airway epithelial cells produce IP-10 (CXCL10) in vitro and in vivo upon rhinovirus infection. Am J Physiol Lung Cell Mol Physiol 2005;289:L85–L95.

30 Lopez-Souza N, Dolganov G, Dubin R, Sachs LA, Sassina L, Sporer H, Yagi S, Schnurr D, Boushey HA, Widdicombe JH: Resistance of differentiated human airway epithelium to infection by rhinovirus. Am J Physiol Lung Cell Mol Physiol 2004;286:L373–L381.

31 Wark PA, Johnston SL, Bucchieri F, Powell R, Puddicombe S, Laza-Stanca V, Holgate ST, Davies DE: Asthmatic bronchial epithelial cells have a deficient innate immune response to infection with rhinovirus. J Exp Med 2005;201:937–947.

32 Contoli M, Message SD, Laza-Stanca V, Edwards MR, Wark PA, Bartlett NW, Kebadze T, Mallia P, Stanciu LA, Parker HL, Slater L, Lewis-Antes A, Kon OM, Holgate ST, Davies DE, Kotenko SV, Papi A, Johnston SL: Role of deficient type III interferon-lambda production in asthma exacerbations. Nat Med 2006;12:1023–1026.

Maria Xatzipsalti
23–27 Makrigianni str.
Makrigianni
GR–Athens, 11742 (Greece)
Tel. +30 210 923 3029, +30 697 445 6750, Fax +30 210 777 6964
E-Mail marax5873@hotmail.co.uk

Sjöbring U, Taylor JD (eds): Models of Exacerbations in Asthma and COPD.
Contrib Microbiol. Basel, Karger, 2007, vol 14, pp 42–67

Modeling Responses to Respiratory House Dust Mite Exposure

*Elizabeth C. Cates, Ramzi Fattouh, Jill R. Johnson,
Alba Llop-Guevara, Manel Jordana*

Department of Pathology and Molecular Medicine, Division of Respiratory Diseases and
Allergy, Centre for Gene Therapeutics, McMaster University, Hamilton, Ont., Canada

Abstract

House dust mite (HDM) is the most pervasive indoor aeroallergen source worldwide.
Allergens derived from HDM are associated with sensitization and allergic asthma. Allergic
asthma is an immunologically driven disease characterized by a Th2-polarized immune
response, eosinophilic inflammation, airway hyperreactivity, and remodeling. Animal models of asthma utilizing ovalbumin (OVA) exposure have afforded us considerable insight with
respect to the mediators and cell types involved in allergic airway inflammation. However,
OVA preparations and HDM are two vastly different materials. This chapter is specifically
concerned with modeling responses to HDM exposure in mice. These studies have furnished
new information and unlocked new lines of inquiry regarding biological responses to common aeroallergens. The complexity of HDM as an allergen source, with its plethora of protein and nonprotein immunogenic components, may influence the mechanisms underlying
sensitization, inflammation and remodeling. Here, we will discuss this issue, along with giving critical thought to the use of experimental models.

Introduction

Allergic asthma is now understood as an immunologically driven disease
that occurs in predisposed individuals as a consequence of aeroallergen exposure. The hallmarks of the asthmatic phenotype principally include the development of a Th2-polarized immune response, eosinophilic inflammation and
airway hyperreactivity (AHR); persistent or recurrent inflammation appears to
be associated with changes in the structure of the airway referred to as remodeling. This chapter is specifically concerned with modeling responses to exposure

to house dust mite (HDM), the most pervasive aeroallergen worldwide. Throughout the chapter, our hope is to place our discussion in the broader context of modeling of asthma, beginning with a foreword about scientific modeling.

Modeling is an essential part of the scientific endeavor. In medicine, illness is manifested by a collection of signs and symptoms. An observer collects them, appraises them and decides to which recognized pattern they best fit. This pattern recognition exercise, essential in deriving a diagnosis, contributes only modestly to the understanding of the origin of a disease, its evolution and its nature. Origin, evolution and nature are facets of a complex process, especially in the case of chronic diseases such as asthma. Advancing our understanding of these facets will come not only from a thorough study of the literature and much careful consideration but, much more likely, through experimentation. Indeed, it has been largely through experimentation in in vitro systems as well as in in vivo animal and human models that we have learned the most about asthma.

Biomedical modeling generally differs from modeling in other scientific domains. In disciplines such as mathematics, physics and economics, a model is the elaboration of a set of rules governing interactions between different variables. Thus, by definition, there is not a predetermined outcome. For example, the equation that defines Newton's Second Law of Motion, $F = ma$ where 'F' is the force applied, 'm' the mass of the object and 'a' is acceleration, is such a model. These kinds of models are particularly suited to explain and predict the outcome of interactions between a number of variables. In sharp contrast, the outcomes of biomedical models are predetermined; in other words, their goal is to recapitulate a given phenotype. The implication of this is that *one can only model what one knows*. This is what determines the logic of biomedical modeling. The undisputed, but somewhat neglected, corollary of this is that modeling of a disease should be a dynamic and ever-evolving endeavor. An underlying theme of this chapter is that modeling promotes questioning which, in turn, echoes the stage of knowledge at a given point and time.

The notion that one can only model what one knows underlies the development of models of asthma over the last 20–30 years. Inflammation, or features of inflammation, had been noted for a long time in asthmatic individuals. However, the appreciation that eosinophilic airway inflammation resulting from a particular type of immunological reaction is a defining feature of the asthmatic phenotype is more contemporary. This understanding, in conjunction with the explosion of molecular immunology, steered the modeling strategies that researchers chose to explore in recapitulating the asthmatic phenotype, particularly in mice. The benefits of this work are apparent as it is now established that allergic asthma is the product of a Th2-polarized response with all of the attendant implications, including the discovery of the functions of archetypical

Th2 cytokines, notably IL-4, IL-5, IL-9 and IL-13, on cardinal features of the asthmatic phenotype.

The vast majority of research in models of asthma to date has utilized chicken egg ovalbumin (OVA), an innocuous protein, as a surrogate allergen. It appears that early research in anaphylaxis led investigators to make this choice, as there is historical evidence that experimental research in anaphylaxis, notably to food antigens, considerably preceded experimental research in asthma. Indeed, studies of OVA-induced anaphylaxis in animal models date back nearly 100 years [1, 2]. Research into these two processes evolved rather separately for a long period of time. It seems plausible that the realization of the conceptual and mechanistic connections between anaphylaxis and asthma beginning with Meltzer in 1910 [3], along with the easy availability and very low cost of OVA, set the stage for researchers wishing to model asthma to settle on OVA as the 'pseudo-allergen' of choice.

Mouse Models of Allergic Airway Disease

OVA-Based Models of Asthma

There have been many studies, particularly over the last 20 years, in experimental models of asthma. The objective of these initiatives has been to recapitulate the cardinal features of the asthmatic phenotype, notably airway inflammation but also AHR and, more recently, airway remodeling. As indicated earlier, most of the research in this field has been performed in models using OVA, a choice that has fundamentally influenced the strategies employed to generate allergic sensitization. Indeed, the fact that exposure to OVA, either intranasally or by aerosol, to naïve mice elicits inhalation tolerance rather than allergic sensitization or airway inflammation [4–7] has forced researchers to develop ways of circumventing inhalation tolerance. With some minor variations, the most commonly used strategy involves the adsorption of OVA to a chemical adjuvant (usually aluminum hydroxide or alum) that is injected into the peritoneal cavity to generate allergic sensitization; we will refer to these models as 'conventional' models of allergic sensitization from here on. An asthma-like phenotype is, then, elicited by a subsequent airway challenge with OVA. The principal utility of these models resides in the fact that they can generate a consistent and robust inflammatory response and, consequently, allow for the study of the characteristics of such a response. Their greatest limitation is that they are inherently designed to chemically bypass the natural requirements for allergic sensitization, precluding the study of that phenomenon. In addition, neglecting the route by which humans are initially exposed to aeroallergens – the respiratory mucosa – may have a number of significant immunological consequences.

That allergic asthma is a result of the interaction between inhaled aeroallergens and the immune system brings to the fore the merit of experimental models capable of generating the asthmatic phenotype by exposure through the respiratory route. It seems intuitive that mucosal exposure to OVA, whether via the respiratory or gastrointestinal mucosa, would result in tolerance for there is neither survival advantage nor immunological parsimony in generating a productive immune-inflammatory response against an innocuous protein. A model system was described in which mice were exposed to aerosolized OVA for 10 consecutive days, leading to their sensitization [8]. However, this has not been the finding in most cases. It seems to follow that the lack of inherent immunogenicity of an antigen can be amended by altering the immune status of the host, particularly through the activation of the surveillance apparatus that first interacts with antigens. For example, studies in which granulocyte macrophage colony-stimulating factor (GM-CSF) was overexpressed in the airway milieu enlighten this point because GM-CSF is one of the most powerful natural adjuvants with substantial effects on the maturation and activation of macrophages and dendritic cells. Indeed, it was shown by Stämpfli et al. [9] that GM-CSF enrichment of the airway microenvironment inhibits tolerance induction to OVA and elicits a phenotype with the typical hallmarks of asthmatic inflammation. Mice subjected to 10 OVA aerosolizations in the context of a GM-CSF-enriched microenvironment showed an expansion and activation of antigen-presenting cells (APCs), developed an OVA-specific Th2-polarized immune response associated with airway eosinophilic inflammation and goblet cell hyperplasia, and exhibited AHR. It is likely that molecules other than GM-CSF, such as thymic stromal lymphopoietin, have a direct role in the generation of allergic immune responses [10, 11].

A series of subsequent studies demonstrated that airway overexpression of several cytokines and chemokines such as IL-12, IL-10 and IP-10 were capable of generating distinct immune-inflammatory phenotypes to the same antigen, OVA [12–14]. Together, these studies illustrate that the *context* in which the immune system first sees antigen is of preeminent relevance to the generation of future immune responses and established experimentally, as stated by Lee and Lee [15], the 'macroimportance of the microenvironment'. Yet this line of experimentation was limited, as it did not account for the fact that common allergens are fundamentally different from OVA, and that humans are rarely ever exposed solely to allergens, or to purified allergenic proteins but rather to complex materials. In other words, the biochemical and presumably immunogenic differences between OVA and the allergenic matter humans inhale, such as HDM, may impact not only the outcome of allergen exposure but also the mechanisms by which these outcomes are brought about.

HDM as Allergen

The discovery of *Dermatophagoides pteronyssinus* as the source of the long-sought 'house dust allergen' in 1967 [16] sparked an exponential growth in investigations into the roles of HDM in allergy [17]. Dust mites are ubiquitous throughout humid areas of the world [18]. Of the phylum Arthropoda and class Arachnida, mites are more closely related to spiders, scorpions and horseshoe crabs than they are to insects, and are therefore difficult to control using routine pest control methods. The small size of many mite species (e.g. 20–320 μm for *D. pteronyssinus*, depending on developmental stage) further complicates control of mite populations, all but ensuring exposure. Thirteen different species of mites can be found in house dust, but the three most common species worldwide, and the major sources of mite allergen, are *D. pteronyssinus*, *D. farinae*, and *Euroglyphus maynei*. In tropical climates, *Blomia tropicalis* is also highly prevalent [19]. The discovery of the allergenic role of mites and an understanding of their biology has allowed researchers to address many questions concerning the global distribution of dust allergies, the seasonality of the disease, and the risk associated with damp houses. Many of these issues are still being investigated today.

It has been argued, based on the global prevalence of mite allergens, and the high proportion of total IgE directed towards them, that there is more specific IgE to mite allergens than any other single allergen source in the world [17]. Sensitization to allergens derived from HDM is also strongly associated with allergic asthma, further emphasizing the clinical importance of this allergen source. Much of the knowledge we have in support of which HDM proteins constitute 'allergens' comes from the binding of patient IgE in Western blotting studies. However, these studies are difficult to interpret due to the variation between reports, the potential for cross-reactivity of antibodies, and the lability of antigens [reviewed in 20]. Therefore, identifying which HDM components are the most universally relevant to human disease, and developing appropriate models in which to test hypotheses surrounding HDM allergy, add additional layers of complexity.

Domestic mites are complex organisms that produce thousands of different proteins and other macromolecules. The identification and isolation of mites from dust and their subsequent culture laid the foundation for the development of the extracts currently used in research. HDM extracts are made from an aqueous solution of a variable mixture of whole mites, nymphs, fecal pellets, eggs and spent culture media, all of which are relevant sources of allergenic proteins [21, 22]. Based on a study of these extracts and the high concentrations of these allergens in dust, it is believed that the group 1 and group 2 HDM allergens are 'the major' allergens. However, it is unknown how well these mite extracts actually compare to the allergens present in the environment. It is possible that some mite proteins present in house dust are poorly represented in extracts due to, for example, their hydrophobicity, stability and/or ability to withstand the extraction process. In addition, extracts differ based on production methods [23], and are difficult to standardize based on the number of proteins they contain [20].

The advent of molecular biology has allowed for the identification, cloning, and synthesis of the proteins identified as HDM allergens. These developments provided the opportunity to investigate the effects of isolated native proteins or recombinant peptides in allergic asthma to HDM. While this approach permits investigation into the effects of a single agent, there are also some shortcomings. First, some recombinant proteins do not retain potentially important native biological activities after being synthesized without a chemical treatment that may affect the protein in other ways. Also, that exogenous adjuvants are required to elicit allergic sensitization to these individual proteins precludes investigation into the immunological mechanisms that underlie this process. In addition, humans are not exposed to a single protein, but to a complex material made up of a multitude of molecules that have the potential to interact. Thus, experiments utilizing a single HDM protein may not capture the complexity of this 'material'. On the other hand, HDM and its extracts are complex; therefore, clarifying the functions of their individual active components and how they contribute to immune responses is a challenging endeavor. This dilemma will likely be mitigated by the convergence of progress arising from the mindful exploration of each experimental paradigm.

Acute HDM Exposure Models

Despite the evidently important contribution of HDM proteins to the myriad of allergens that humans are exposed to, relatively little research has been conducted in animal models to discern the nature of the effects of HDM in vivo. To date, the vast majority of this work has been done using the previously mentioned 'conventional' [intraperitoneal (i.p.) + alum] models of allergic sensitization. Shortly following the discovery of mites as the allergenic source of house dust, and the establishment of culture systems from which to extract mite proteins, research logically focused on the ability to elicit IgE responses in mice and guinea pigs [24–27]. Initially it was discovered that it is possible to sensitize guinea pigs to HDM extracts via the respiratory mucosa [26]. At this time however, sensitization in the absence of adjuvant appeared to be arduous, taking as long as 17 weeks to achieve [24]. In mice, multiple injections of HDM seemed to be required, even in the presence of alum, to generate an IgE response [27, 28]. These early findings appear to have set the stage for the primary use of conventional models of allergic sensitization as the standard in the field. This is true for the majority of studies that have followed, modeling both allergic asthma and atopic dermatitis, unless modified animals such as T cell receptor transgenics [29] were employed.

Research investigating the immunological mechanisms involved in allergy to HDM began in earnest in the late 1980s to early 1990s. Experiments addressing the allergenic components of HDM extracts have demonstrated that while the higher molecular weight fraction of these extracts (containing Der p 1) does appear to retain the majority of the allergenic potential [30], Der p 1 alone induces oral and inhalation tolerance [31–33]. It is plausible that this finding perpetuated the notion that conventional models were needed to examine the induction of responses to HDM, and the continued use of i.p. models sparked two lines of questioning. One line is concerned with the development models of mucosal sensitization that are more clinically relevant [29, 34, 35], and the other asks immunological questions using conventional i.p. models and the inflammatory responses they elicit. Studies in this latter category have been useful in addressing questions such as the role of eosinophils in AHR elicited by HDM [36], the types of inflammatory responses elicited by different species of mites [37], the role of the proteolytic activity of Der p 1 in inflammatory responses in the lung [38, 39], the importance of the spatial distribution of immune cells in the lungs of primates [40], as well as the feasibility of cytokine gene therapy-based strategies or parasitic infections to treat HDM allergy [41, 42].

As previously mentioned, early attempts to sensitize mice to HDM via the respiratory mucosa were met with limited and inconsistent success. In 1999, Yu et al. [43] demonstrated that sensitization to a crude *D. farinae* extract was possible via the intratracheal route in the absence of exogenous adjuvants. However, the inflammatory responses in the lung and IgE titers obtained were admittedly low, and much more robust responses could be achieved via conventional i.p. sensitization followed by airway challenge. This was followed in 2004 by Cates et al. [44] who published an acute model of allergic sensitization to an HDM extract (*D. pteronyssinus*) in which mice were indeed sensitized via the respiratory mucosa in the absence of exogenous adjuvants. This model was characterized by the current hallmarks of allergic asthma: the expansion and activation of APCs, an eosinophilic inflammation in the lung, deployment of a Th2-polarized immune response, Th2-associated immunopathology and resultant AHR. This acute HDM exposure model also allowed for a preliminary investigation into the roles of a cytokine, GM-CSF, in the early stages of allergic sensitization. In an interesting parallel to the previously reported findings of Stämpfli et al. [9], who demonstrated that enriching the airway environment with GM-CSF allowed for sensitization to OVA, the administration of an anti-GM-CSF antibody in the Cates et al. study markedly decreased responses to HDM. This was believed to be due to the ability of HDM to elicit GM-CSF production from bronchial epithelial cells [45] and the powerful effects this cytokine has on APCs [reviewed in 46]. Therefore, unlike OVA, HDM appeared to have an intrinsic ability to elicit an immune response, the nature of which can now be studied in more detail. Using a slightly modified model, others later reported similar findings, further expanding their work to examine the impact of regulatory T cells on dendritic cells and allergic responses to HDM [47].

One of the fundamental differences between the models discussed above and allergen exposure in humans is that these animals were exposed to HDM antigens for short, experimentally defined periods of time, whereas human exposure is a ubiquitous life-long event. As such, these kinds of models are inadequate to examine questions surrounding the biological processes that occur as a result of chronic allergen exposure, and if or how these processes play a role in diseases such as allergic asthma.

Modeling the Chronic Aspects of Allergic Asthma

That human exposure to HDM is a chronic event implies that there is a persistent, ongoing interaction between HDM antigens and the immune system. However, the details of the consequences of this interaction in allergic and non-allergic persons are unknown. Currently, it is believed that chronic exposure to allergen results in persistent eosinophilic inflammation, which leads to the structural changes, characteristic of asthma, known as airway remodeling.

Remodeling of the airways was first described quantitatively in 1922 by Huber and Koessler [48], and many subsequent studies have shown this phenomenon to be comprised of increases in mucous production, subepithelial extracellular matrix deposition, and airway contractile elements such as airway smooth muscle cells and myofibroblasts. As the clinical significance of these structural changes to the airway wall has become apparent, efforts have been made to develop model systems to investigate immunological links to this phenomenon.

Many cell types and mediators associated with Th2-polarized immune-inflammatory responses have been implicated, primarily in vitro, in the elaboration of airway structural changes. Cytokines such as IL-13 and IL-4 have been shown to directly effect airway structural cells, leading to increased mucous production [49] and collagen deposition [50], as well as phenotypic changes in fibroblasts [51]. Activated Th2-polarized lymphocytes have also been directly implicated in the increased proliferative capacity of airway smooth muscle cells [52]. Moreover, recent studies into the structural consequences of eosinophil degranulation in the airway have revealed that eosinophil-derived cationic proteins are able to modify gene expression in bronchial epithelial cells, inducing them to synthesize numerous mediators, such as growth factors and matrix metalloproteinases, which ultimately lead to altered extracellular matrix composition and turnover [53]. In addition to the direct impact of inflammatory cells on airway structure, resident structural cells themselves, including epithelial cells, smooth muscle cells, fibroblasts and myofibroblasts [reviewed in 54], may participate in the initiation of a repair phenotype characterized by increased collagen deposition, which is thought to be driven by TGF-β and other growth factors [55]. Due to the complexity of remodeling, and the number of cell types and mediators involved, the extent and nature of the contributions made by inflammatory and structural cells in airway remodeling and the asthmatic phenotype as a result of chronic allergen exposure in vivo remains to be elucidated.

To address these questions, significant effort has been exerted to recapitulate the chronic aspects of asthma, using animal model systems. One approach to investigating the role of Th2-associated molecules in airway remodeling has been the genetic manipulation of the host. By either upregulating pulmonary expression of key Th2 cytokines such as IL-9 and IL-13 [56, 57] or by knocking out expression of t-bet [58], a transcription factor associated with the generation of Th1 immune responses, chronic airway inflammation, peribronchial fibrosis and abnormal airway function can be established. While these models illuminate the potential role of each of these factors in the structural changes associated with asthma, they ignore the potential contribution of allergen exposure itself on the development and maintenance of abnormal airway structure and function.

Acute protocols employing short-term OVA delivery, even at a high dose, have been used to examine the impact of allergen exposure on airway remodeling. However, these protocols do not lead to most of the structural changes associated with chronic asthma. Therefore, the effects of chronic pulmonary exposure to OVA were explored. Interestingly, exposure under these conditions led to an abrogation of the immune-inflammatory response to OVA. Although initial delivery of OVA results in immune activation, sustained respiratory exposure leads to an attenuation of the immune response to OVA, deviation of the immunoglobulin profile from IgE to IgG1, decreased expression of costimulatory molecules on lung mononuclear cells, and decreased eosinophilia in the peripheral blood and bronchioalveolar lavage, even in i.p. sensitized animals [59].

In order to avoid the development of inhalation tolerance to OVA, various models have been developed in which OVA is delivered intermittently after i.p. sensitization [60–63]. The resultant phenotype of these models includes epithelial lesions (goblet cell hyperplasia, mucous production and epithelial shedding) and alterations to the subepithelium, most notably increased collagen deposition and accumulation of contractile elements (smooth muscle hypertrophy/hyperplasia and myofibroblast differentiation). Many of these models are also associated with increased AHR to methacholine. Interestingly, the extent of eosinophilia subsides progressively with each exposure. In addition, interactions between many common aeroallergens and the lung microenvironment are a potentially important pathway in the development of altered airway wall structure. This is an aspect that cannot be addressed using OVA as the antigen.

To this effect, models employing chronic exposure to HDM were developed. Recently, in the absence of any adjuvant, chronic exposure to a *D. farinae* extract has been reported to induce airway eosinophilia in infant monkeys [64]. Subsequently, in 2004, Johnson et al. [65] described a mouse model of sustained allergic airway inflammation and remodeling employing chronic, sustained respiratory delivery of a *D. pteronyssinus* extract. This protocol of chronic respiratory exposure to HDM led to sustained airway eosinophilic inflammation, which reached a plateau following 3 weeks of exposure. HDM delivery was associated with immune activation in the lung, characterized by the expansion of APC, activated T cell, and Th2 effector cell populations following 5 weeks of continuous HDM exposure. An allergen-specific memory response to HDM extract was also observed by elevated in vitro production of IL-4, IL-5 and IL-13 by splenocytes isolated from mice chronically exposed to HDM, and humoral immunity to HDM was indicated by elevated serum levels of IgE and HDM-specific IgG_1.

This chronic eosinophilic immune-inflammatory response was also accompanied by profound changes to the structure of the large airways. In addition to the goblet cell hyperplasia and increased mucous production commonly observed in acute allergen exposure protocols, continuous HDM exposure resulted in an increased expansion of peribronchial contractile elements, presumably myofibroblasts, as well as collagen deposition in the airway wall subepithelium. Finally, chronic exposure to HDM had profound functional consequences, with markedly increased AHR to methacholine. Interestingly, the lung structural changes and dysfunction remained following HDM withdrawal, despite an attenuation of the inflammatory response. This suggests that the airway dysfunction seen in this model is contributed to by both airway inflammation and remodeling. It is hoped that a detailed exploration of the evolution of the structural and functional abnormalities in this model will help to improve our understanding of the mechanisms contributing to airway remodeling and dysfunction in the context of chronic exposure to a common aeroallergen.

Considering the Allergenicity of HDM

Clearly, allergic sensitization to a variety of allergens, including OVA and HDM, can be achieved experimentally. These systems do not inform matters such as: (1) what makes allergens like HDM allergenic and (2) under what conditions allergic sensitization naturally occurs. Confronting such questions is the selective pressure that has driven past changes in modeling and will drive future lines of inquiry.

The Biochemical Nature of HDM Extracts

As previously mentioned, OVA preparations and HDM extracts are, from the perspective of biochemical diversity, two vastly different materials. The OVA preparations that are commonly used in models of allergic disease are typically >98% pure and hence, are essentially free from additional protein and nonprotein components. In stark contrast, HDM extracts are complex materials consisting of hundreds of protein and nonprotein components that range in both size and function. Indeed, it is this fundamental difference between OVA and HDM that is thought to account for the divergent immunological outcomes which arise following inhalation of these materials. This realization has fueled investigations aimed at characterizing HDM-derived allergens and understanding the mechanisms by which these individual components promote allergic disease.

It is estimated that mites produce approximately 3,000 proteins, 5% of which are thought to be allergens [66]. Studies conducted in the mid-late 1980s documented >30 IgE-binding proteins in dust mite extracts using sera from mite-allergic patients [67–69]; however, the actual number of *different* mite allergens represented in those studies may be somewhat less. According to the listing maintained by the Allergen Nomenclature Sub-Committee of the International Union of Immunological Societies [70], 21 different groups of dust mite allergens have been identified. A biological function has been described for most of these allergens, and collectively, they encompass a broad range of functions including binding proteins, structural proteins, enzyme inhibitors and various enzymes [thoroughly reviewed in 20].

The majority of investigations into the impact of HDM-derived allergens on the pathogenesis of allergic disease have focused on the proteolytic activity of HDM proteins. There are at least 4 HDM allergens that possess proteolytic activity. The group 1 allergens display a mixed cysteine/serine protease activity [71], while groups 3, 6 and 9 are serine proteases [72–74] and a number of mechanisms have been identified by which the proteolytic activity of these allergens is thought to contribute to HDM's allergenicity.

A series of in vitro studies conducted by Wan et al. [75, 76] using solubilized Der p 1 demonstrated that this protease disturbed a monolayer of cultured bronchial epithelial cells, resulting in minor exfoliation and the disruption of tight junctions. Interestingly, this event was associated with increased epithelial cell permeability and transepithelial migration of Der p 1. Similar effects were observed when bronchial epithelial cells were treated with culture medium enriched in serine proteases derived from *D. pteronyssinus* (i.e. rich in Der p 3, 6 and 9) [77]. Together, these results suggest that the proteolytic activity of HDM may facilitate allergen entry into the submucosal environment. Consequently, this may increase HDM's contact with APCs and the likelihood of developing allergic sensitization [76].

In addition to enhanced epithelial permeability, it has become increasingly clear that the proteolytic activity of HDM-derived allergens can have more direct proinflammatory effects. Alveolar macrophages treated with a *D. farinae* extract in vitro respond by producing IL-6, TNF-α and nitric oxide [78]. Similarly, purified Der p 1, 3 and 9 were collectively shown to induce the production of GM-CSF, IL-6, IL-8 and eotaxin from bronchial epithelial cell lines and primary cultures of human bronchial epithelium [45, 79]. Although the precise mechanism of cytokine induction from mite allergen-treated epithelial cells differs depending on the allergen and cytokine, for the most part these responses seem to be mediated by the activation of protease-activated receptor-2 [79, 80]. More recently, a protease-independent mechanism of Der p 1-mediated cytokine release has been reported [81, 82]. Notably, an HDM allergen lacking any known proteolytic activity (Der p 5) was shown to induce cytokine expression from epithelial cells [82]. Taken together, the increased accessibility of allergens resulting from disrupted epithelial integrity, and a proinflammatory microenvironment may strongly promote the development of allergic sensitization. However, it is important to note that the extent to which these events occur in vivo remains unclear.

It has also been suggested that HDM's proteolytic activity may enhance IgE synthesis as Der p 1, in vitro, can cleave the low-affinity IgE Fc receptor (CD23) from the surface of human B cell lines [83, 84]. CD23 is a key regulator of the IgE network as its binding of IgE and IgE-containing immune complexes on the surface of B cells triggers a negative-feedback signal that prevents further IgE synthesis [85, 86]. Moreover, soluble CD23 has been previously reported to enhance IgE synthesis [87–89]. Thus, it has been proposed that Der p 1-mediated cleavage of CD23 may not only impair IgE feedback inhibition pathways but may also augment IgE synthesis directly [83]. Additionally, both HDM extracts and purified Der p 1 can induce the production of IL-4 directly from mast cells and basophils [90], and this may also enhance the IgE response.

The proteolytic activity of dust mite allergens may, furthermore, bias the development of a Th2 response by cleaving the α-subunit of the IL-2 receptor (CD25) on T cells [91]. Schulz et al. [91] have demonstrated that Der p 1-mediated cleavage of CD25 from human peripheral blood lymphocytes resulted in decreased proliferative and cytokine (IFN-γ) responses upon stimulation with anti-CD3 antibody in vitro. Subsequently they demonstrated that Der p 1-conditioned human CD4 and CD8 T cells produce more IL-4, and less IFN-γ, following stimulation with anti-CD3 in vitro than T cells treated with inactivated Der p 1 [92]. These findings suggested that HDM may limit the development of Th1 response, at the same time promoting the development of a Th2 response and IgE synthesis [91]. Importantly, this notion is supported by in vivo evidence showing that mice immunized with proteolytically active Der p 1 generate higher titers of Der p 1-specific IgE compared to controls immunized with inactivated Der p 1 [93].

The recent discovery of the involvement of chitinases in the pathogenesis of experimentally induced allergic airway inflammation may have uncovered another mechanism by which the functions inherent to HDM proteins may contribute to their allergenicity [94]. Chitinases are enzymes that catalyze the hydrolysis of chitin and are found in arthropods such as dust mites. In addition, a family of chitinase and chitinase-like genes has been identified in both rodents and humans [95–97].

Elias et al. [98] have postulated that some mammalian chitinases may act as potent mediators of the Th2 immune response. In support of this concept, they found increased expression of acidic mammalian chitinase (AMCase) following sensitization and challenge with OVA and, importantly, neutralization of AMCase markedly reduced inflammation and AHR [94]. The expression of mammalian chitinase-like proteins has also been observed in the context of Th2-mediated inflammation during experimental parasitic infection [99]. Moreover, a putative member of the chitinase protein family, ECF-L, may be a potent chemoattractant for eosinophils and T cells [99, 100]. These findings indicate that mammalian chitinases/chitinase-like proteins may play a role in the modulation of the allergic response.

The group 15 and 18 HDM allergens have been identified as chitinases [101, 102]. Based on the findings discussed above, it is reasonable to speculate that the HDM-derived chitinase allergens may function in a manner similar to human chitinase proteins to potentiate Th2 responses. To our knowledge, there is no direct evidence in support of this. However, the group 15 and 18 dust mite allergens show a high frequency of binding to IgE in patient sera, and are the main IgE-binding proteins in allergic dogs [102–104]. This supports the notion that HDM-derived chitinase allergens may influence allergic disease.

Clearly, the functional properties of allergens can profoundly influence their ability to elicit an allergic immune response, although it is likely that in some cases function may be of little significance. For the most part, as it pertains to many of the HDM-derived allergens, this information is simply lacking at this time. For example, the group 10 and 11 allergens have been identified as muscle proteins (tropomyosin and paramyosin, respectively) and have been shown to bind IgE at a high frequency [105, 106]. Similarly, the group 2 allergens bind a high proportion of patient IgE, and have thus been identified as major allergens. In addition, the group 14 allergens are abundant mite proteins that function as lipid transport or storage molecules [107, 108]. Interestingly, they have demonstrated both potent reactivity in T cell assays and high IgE binding. It remains to be seen if these biochemical functions or immunological activities contribute to HDM's allergenicity.

The complexity of HDM extracts is further enhanced because, in addition to the diverse array of protein allergens, they contain a number of nonprotein components. Of these, lipopolysaccharide (LPS) is the best defined. It is widely acknowledged that LPS is a potent stimulator of the innate immune response and, correspondingly, has powerful adjuvant activity. LPS can elicit the production of a panopy of proinflammatory mediators including cytokines such as TNF-α, GM-CSF, IL-6, IL-8, IL-10 and IL-12 [109, 110], some of which are thought to contribute to the development of allergic responses [110, 111]. That LPS exposure can facilitate allergic sensitization and exacerbate airway inflammation has been demonstrated in experimental systems [112, 113] and moreover, is associated with the severity of allergic airway diseases in humans [110]. Numerous other nonprotein immunostimulators such as lipoteichoic acid and peptidoglycans are abundantly present in the environment, and at least some of these may promote allergic sensitization [114]. However, aside from LPS, very little is known regarding the content of such molecules in HDM. Given their ubiquitous distribution and profound influence on innate immunity, it is likely that at least some of these molecules may also contribute to HDM's inherent allergenicity. Clearly, a comprehensive description of the nonprotein constituents of HDM is required.

It has become apparent that HDM extracts possess, in sharp contrast to OVA, a number of entities capable of distinctly engaging the immune system. What then, could be the immunological relevance of the biochemical differences between OVA and HDM? It seems intuitive that the biochemical and immunogenic profile of HDM extracts will draw on an expanded network of cellular responses and unique molecular signatures. Take for example, the outcome of airway remodeling. While OVA can, under certain conditions (e.g. i.p. sensitization with intermittent exposure) lead to remodeling, achieving the same outcome utilizing an HDM extract and a route of mucosal sensitization does not imply the same underlying biochemical or immunological process; nor can this knowledge be universally applied to the responses elicited by other aeroallergen families. In this regard, preliminary data from our laboratory suggests that in contrast to OVA-based systems [115], TGF-β may *not* be involved in the generation of HDM-induced airway remodeling [116]. So, while some of the experimental outcomes of using OVA or HDM may be quite similar, the mechanistic processes that underlie those outcomes are likely to be fundamentally different.

The Issue of Exposure

A common denominator of the studies cited above is that, in all instances, the concentration of HDM extract administered was experimentally determined by the researchers to achieve their desired outcome (e.g. 25 μg/day/mouse). Therefore, a legitimate question is: how comparable is this concentration of allergen to human exposure? Absolutely the question is intuitive and legitimate; however, the answer is just as elusive. Arguably, the significance of this issue is relative, because *a model of a chronic disease such as allergic asthma is the compression in time and space of a protracted process.* 'In time' because the objective is to achieve a predetermined outcome in a relatively short period of time; indeed, the alternative of achieving such a goal in 1–2 years would be a self-defeating enterprise. '*In space*' because, clearly, 'real life' is not the same as life in a cage in an animal facility or, for that matter, the environment in a university research laboratory (in the case of human research). These are the inherent limitations of experimental modeling where the trade-off between 'real life' context and the control of variables and productivity must be constantly scrutinized. The relevance of the question of comparative exposure is also dependent on the nature of the overall research question being asked. If the query is: what concentration of, and under what conditions, an HDM extract elicits allergic sensitization and/or airway inflammation, then the information derived from using a single concentration of 25 μg is largely uninformative. Studies comprehensively examining a wide range of doses and lengths of exposure would shed light on these questions. However, if the issue is to consistently establish a robust phenotype with the typical hallmarks of asthma, then, that concentration is suitable. Nevertheless, these comments should not be construed as intellectual trickery to circumvent the genuine relevance of the initial question: how comparable is this experimental dose of allergen to human exposure? The reply is saturated with uncertainty because the terms of reference are ambiguous. Indeed, many clinical and epidemiological studies have measured concentrations of mite allergens. However, the site sampled, the sampling techniques used (type of collectors, frequency of sampling, environment), and the methods of measurement and quantification of allergen employed are diverse. Moreover, the criteria employed to define 'risk' (for sensitization as well as disease) are not uniform. Therefore, to precisely identify the concentrations of HDM associated with allergic sensitization and airway disease in humans is an arduous endeavor.

With respect to allergen sampling, there is still debate as to whether *indirect* personal exposure should be assessed by measuring allergen, usually Der p 1, either in airborne dust, in settled dust, or in allergen reservoirs [117]. *Airborne allergen* has been commonly measured by air samplers or ion-charging devices [117, 118]. Under these conditions, allergen is usually expressed as nanograms per cubic meter of air. Most investigators have been unable to detect airborne mite allergens in the absence of experimental or household disturbance, likely because these levels are below the detection limit for most assays. Allergens in *settled dust* can be detected and measured by leaving Petri dishes out for 2 weeks or using an adhesive-membrane system [117]. In this case, allergen-settling rate is calculated as nanograms per square meter per day. A third method used to measure allergen concentrations is to sample *allergen reservoirs* by means of vacuum sampling [117]. In this instance, exposure to indoor allergens is usually expressed as the concentration of allergen detected in dust samples (μg/g dust); this is considered to be the most consistent method available [119]. Despite controversy as to how the allergen concentrations measured using these varied sampling methods might correlate with each other [117, 120], measurement of mite allergen concentration in dust reservoirs, as a surrogate for airborne exposure, is the most commonly used method.

The site of allergen sampling is also relevant. Mattresses, carpets/floors and upholstered furniture are the best places indoors to detect mites and mite products. Other places with adequate textile substratum and optimal temperature, humidity, and food resources for mites are shutters, clothing and picture frames [121]. It has been proposed that mattress-dust sampling is the most appropriate marker of indoor allergen exposure because bedding is a major reservoir of mite allergens and probably the most important environment for exposure. A number of studies have documented that the amount of mite allergens found in mattresses is up to 10-fold higher than in other areas of a household [122, 123]. It should be noted that the distribution of allergen is dynamic. In particular, the concentration of airborne allergen is notably influenced by any 'disturbance factor'. For example, it has been shown that the airborne levels of Der p 1 increase about 1,000-fold in disturbed conditions, such as bed making [122] or vacuum cleaning done without a filter.

With these considerations in mind, many clinical and epidemiological studies have examined the amount of mite allergen present in homes. The central message that emerges from these studies is that there is an extraordinary geographical variability in terms of both the number of households where mites can be detected and the concentration of mite proteins detected at each site. A recent study by Zock et al. [124] examined over 3,500 households in 22 European cities representing 10 countries. The percentage of households in which mite allergens were detected ranged from 0 to 98.4% (globally, an average of 48.4%). Detection of mite allergens was performed by vacuuming 1 m^2 of bed for 2 min; under these conditions, the concentration of mite allergens detected (Der p 1 and Der f 1) ranged from 0.01 to 14.71 μg/g of dust. Other European studies have found concentrations of mite allergens ranging from 0.18 μg/g of dust (carpets) and 5.6 μg/g of dust (mattresses) in Germany [125], to 16 and 28 µg/g of dust in Southern England and Manchester [126, 127], respectively. Leaderer et al. [128] showed that 47% of households in Connecticut and Central Massachusetts had levels of HDM greater than or equal to 2 µg/g of dust, and in 22% of households levels were greater than or equal to 10 µg/g of dust. Higher levels were reported in coastal Australia and New Zealand where prevalence and severity of asthma is the highest [129]; indeed, a study in Wellington reported that at least 2 µg/g of dust were detected in all households, and 36% had levels greater than 100 µg/g of dust [130, 131]. Finally, many studies have shown that there is a plethora of variables influencing the concentration of mite allergens in the household including humidity in particular, temperature, living conditions such as dampness, number of people living in a household and availability of central heating, living habits such as vacuuming mattresses and sleeping with the window open in winter, and many others [119, 124, 132–134]. Some of these variables not only influence the amount of mite allergen that can be detected but also they are likely to alter the actual amount of airborne allergen.

It is, at this point, tempting to furnish an answer to the question of how daily human exposure to mite allergens compares to the daily exposure of mice to these same allergens in experimental systems. The amount of HDM that interacts with the respiratory mucosa of the lower airway of a mouse may be 5- to 10-fold less the amount deposited into its nostrils [135]. Despite this degree of fluctuation, the actual amount of HDM a mouse is experimentally exposed to per day can be reasonably estimated. Clearly, it is far less direct to deduce the amount of allergen interacting with the lower airway respiratory mucosa of a human on a per day basis from the number of micrograms of Der p 1 per gram of collected dust. For example, that several micrograms of Der p 1 can be measured in a sample that has been collected by vacuuming 1 m² of a mattress for 2 min says little about how much was made airborne, inspired, and passed the necessary physical barriers to reach the lower respiratory tract. In other words, the logical ambiguity in precisely determining mucosal exposure to HDM in humans precludes the beneficial intent to formulate a rigorous interspecies comparison of exposures.

A different but relevant issue is the relationship between levels of exposure and the risk for sensitization, an altogether different issue to that of the relationship between allergen exposure and allergic disease. While some controversy remains, the consensus is that there is an association between measurable levels of HDM allergen and prevalence and severity of allergic sensitization [136–138]. However, whether this relationship is linear remains to be elucidated. Different studies have provided somewhat different estimates, but it has been put forth that Der p 1 levels as low as 2 µg/g of dust are a significant risk for allergic sensitization [119, 139]. While such measurements are clearly useful as an epidemiological reference in defined settings, extrapolation of detected amounts of Der p 1 in dust to explaining pathogenesis may extend beyond applicability. That some studies have failed to uncover a relationship between the concentration of mite allergens per gram of dust and the extent of allergic sensitization is not highly surprising, as there may be a number of reasons accounting for this apparent disconnect. For example, humans are not really exposed only to Der p 1, or Der f 1, but to complex materials, which, as previously discussed, contain other immunologically active moieties. Thus, the mite materials humans are exposed to 'in the wild' are more than likely to be rather heterogenous. In addition, immune responsiveness to HDM (or to Der p 1) is surely influenced by the immune status of the host at the time of exposure. For example, concurrent exposure to other immunologically active entities such as molds and viruses, preexisting sensitization to other non-HDM allergens, or exposure to pollution among other things may prime the immune system altering, either way, the threshold of responsiveness to mite allergens. This complex interplay would directly impact on the concentration of HDM (or Der p 1) required to elicit allergic sensitization. In any case, failure to find a correlation between allergen exposure and sensitization does not justify the conclusion that these two variables are not related.

Closing Remarks

Evidently, modeling is pivotal to advancing knowledge in any scientific discipline. In the medical sciences, experimental modeling aims to recapitulate a predetermined outcome. While the generally implied goal of modeling is to 'mimic a disease', we must be mindful that this is a misguided enterprise. Overall, disease is an experience that is virtually impossible to duplicate, especially across species. Indeed, the experience and manifestation of disease have not only many anatomical and physiological constraints but also fundamentally distinct social and psychological aspects, all of which have an impact on the immune system. Perhaps, the most constructive way to conceptualize modeling is that it is an attempt to recapitulate the phenotype that we think, *based on our best understanding at a given point in time*, underlies a disease state.

Indeed, the notion of '*our best understanding*' does not refer only to the amount of information available at any point in time. Of course information plays a central role but equally, if not more important, is the current conceptualization, the established paradigm that 'explains' the disease. This, we surmise, is what fuels future and novel modeling initiatives. In other words, if asthma is understood, as it was historically, as a 'nervous imbalance', or a 'mediator hyper-release process', then modeling will go along one path. If, in contrast, asthma is understood as an immune-driven disease, then modeling will evolve along a very different path, both practically and conceptually.

A number of experimental models of asthma have been developed to date. The past, present and future usefulness of each one of them rests in the ability of researchers to remain both cognizant of the limitations of each experimental construct, and open to the changes that must occur with time and discovery. The greatest liability of exploiting a model uncritically is that it may end up furnishing knowledge alienated from, and hence of doubtful relevance to, the very disease process it was meant to investigate.

Acknowledgements

The authors wish to thank Amal Al-Garawi and Alexander Caudarella for helpful discussion. ECC and RF were the holders of Canada Graduate Scholarships from the CIHR and ECC is the holder of an Eva Eugenia Lillian Cope Research Scholarship. JRJ is the holder a CIHR Doctoral Research Award, AL-G holds a La Caixa Scholarship (Spain) and MJ holds a Senior Canada Research Chair in Immunobiology of Respiratory Diseases and Allergy.

References

1　Wells HG: Studies on the chemistry of anaphylaxis. III. Experiments with isolated proteins, especially those of the hen's egg. J Infect Dis 1911;9:147.

2　Laroche G, Richet C, Saint-Girons F: Anaphylaxie et immunité alimentaires expérimentales, à l'ovalbumine. C R Soc Biol 1913;1:87–89.

3　McFadden ER Jr: A century of asthma. Am J Respir Crit Care Med 2004;170:215–221.

4　Holt PG, Batty JE, Turner KJ: Inhibition of specific IgE responses in mice by pre-exposure to inhaled antigen. Immunology 1981;42:409–417.

5　Swirski FK, Gajewska BU, Alvarez D, Ritz SA, Cundall MJ, Cates EC, Coyle AJ, Gutierrez-Ramos JC, Inman MD, Jordana M, Stämpfli MR: Inhalation of a harmless antigen (ovalbumin) elicits immune activation but divergent immunoglobulin and cytokine activities in mice. Clin Exp Allergy 2002;32:411–421.

6　Tsitoura DC, Blumenthal RL, Berry G, Dekruyff RH, Umetsu DT: Mechanisms preventing allergen-induced airways hyperreactivity: role of tolerance and immune deviation. J Allergy Clin Immunol 2000;106:239–246.

7　Tsitoura DC, DeKruyff RH, Lamb JR, Umetsu DT: Intranasal exposure to protein antigen induces immunological tolerance mediated by functionally disabled CD4+ T cells. J Immunol 1999;163:2592–2600.

8　Hamelmann E, Oshiba A, Loader J, Larsen GL, Gleich G, Lee J, Gelfand EW: Antiinterleukin-5 antibody prevents airway hyperresponsiveness in a murine model of airway sensitization. Am J Respir Crit Care Med 1997;155:819–825.

9　Stämpfli MR, Wiley RE, Neigh GS, Gajewska BU, Lei XF, Snider DP, Xing Z, Jordana M: GM-CSF transgene expression in the airway allows aerosolized ovalbumin to induce allergic sensitization in mice. J Clin Invest 1998;102:1704–1714.

10　Al-Shami A, Spolski R, Kelly J, Keane-Myers A, Leonard WJ: A role for TSLP in the development of inflammation in an asthma model. J Exp Med 2005;202:829–839.

11   Zhou B, Comeau MR, De Smedt T, Liggitt HD, Dahl ME, Lewis DB, Gyarmati D, Aye T, Campbell DJ, Ziegler SF: Thymic stromal lymphopoietin as a key initiator of allergic airway inflammation in mice. Nat Immunol 2005;6:1047–1053.

12   Stämpfli MR, Scott Neigh G, Wiley RE, Cwiartka M, Ritz SA, Hitt MM, Xing Z, Jordana M: Regulation of allergic mucosal sensitization by interleukin-12 gene transfer to the airway. Am J Respir Cell Mol Biol 1999;21:317–326.

13   Stämpfli MR, Cwiartka M, Gajewska BU, Alvarez D, Ritz SA, Inman MD, Xing Z, Jordana M: Interleukin-10 gene transfer to the airway regulates allergic mucosal sensitization in mice. Am J Respir Cell Mol Biol 1999;21:586–596.

14   Wiley R, Palmer K, Gajewska B, Stämpfli M, Alvarez D, Coyle A, Gutierrez-Ramos J, Jordana M: Expression of the Th1 chemokine IFN-gamma-inducible protein 10 in the airway alters mucosal allergic sensitization in mice. J Immunol 2001;166:2750–2759.

15   Lee NA, Lee JJ: The macroimportance of the pulmonary immune microenvironment. Am J Respir Cell Mol Biol 1999;21:298–302.

16   Voorhorst R, Spieksma FTM, Varekamp H, Leupen MJ, Lyklema AW: The house-dust mite *(Dermatophagoides pteronyssinus)* and the allergens it produces. Identity with the house-dust allergen. J Allergy 1967;39:325–339.

17   Tovey E: Effect of Voorhorst's work on the current understanding of the role of house dust mites in allergic diseases. J Allergy Clin Immunol 2004;113:577–580.

18   Arlian LG: Mites are ubiquitous: are mite allergens, too? Ann Allergy Asthma Immunol 2000;85:161–163.

19   Arlian LG, Platts-Mills TA: The biology of dust mites and the remediation of mite allergens in allergic disease. J Allergy Clin Immunol 2001;107:S406–S413.

20   Thomas WR, Smith WA, Hales BJ, Mills KL, O'Brien RM: Characterization and immunobiology of house dust mite allergens. Int Arch Allergy Immunol 2002;129:1–18.

21   Stewart GA, Butcher A, Lees K, Ackland J: Immunochemical and enzymatic analyses of extracts of the house dust mite *Dermatophagoides pteronyssinus*. J Allergy Clin Immunol 1986;77:14–24.

22   Tovey ER, Chapman MD, Platts-Mills TA: Mite faeces are a major source of house dust allergens. Nature 1981;289:592–593.

23   Geissler W, Maasch HJ, Winter G, Wahl R: Kinetics of allergen release from house dust mite *Dermatophagoides pteronyssinus*. J Allergy Clin Immunol 1986;77:24–31.

24   Ishii A, Matsuhashi T, Miyamoto J, Sasa M: Production of homocytotropic antibody in guinea pigs inhalated with an extract of the house dust mite, *Dermatophagoides farinae*. Jpn J Exp Med 1973;43:25–32.

25   Chopra SL, Vincent L: Investigations on alum-precipitated pyridine-extracted allergens. II. Sensitization studies in guinea pigs. Int Arch Allergy Appl Immunol 1974;46:321–330.

26   Ishii A, Takaoka M, Matsui Y: Production of guinea pig reaginic antibody against the house dust mite extract, *Dermatophagoides pteronyssinus*, without adjuvant. Jpn J Exp Med 1977;47: 377–383.

27   Kudo K, Okudaira H, Miyamoto T, Nakagawa T, Horiuchi Y: IgE antibody response to mite antigen in the mouse. Suppression of an established IgE antibody response by chemically modified antigen. J Allergy Clin Immunol 1978;61:1–9.

28   Inagaki N, Tsuruoka N, Goto S, Matsuyama T, Daikoku M, Nagai H, Koda A: Immunoglobulin E antibody production against house dust mite, *Dermatophagoides farinae*, in mice. J Pharmacobiodyn 1985;8:958–963.

29   Jarman ER, Tan KA, Lamb JR: Transgenic mice expressing the T cell antigen receptor specific for an immunodominant epitope of a major allergen of house dust mite develop an asthmatic phenotype on exposure of the airways to allergen. Clin Exp Allergy 2005;35:960–969.

30   Tategaki A, Kawamoto S, Okuda T, Aki T, Yasueda H, Suzuki O, Ono K, Shigeta S: A high-molecular-weight mite antigen (HM1) fraction aggravates airway hyperresponsiveness of allergic mice to house dusts and whole mite cultures. Int Arch Allergy Immunol 2002;129:204–211.

31   Hoyne GF, Callow MG, Kuhlman J, Thomas WR: T-cell lymphokine response to orally administered proteins during priming and unresponsiveness. Immunology 1993;78:534–540.

32   Hoyne GF, O'Hehir RE, Wraith DC, Thomas WR, Lamb JR: Inhibition of T cell and antibody responses to house dust mite allergen by inhalation of the dominant T cell epitope in naive and sensitized mice. J Exp Med 1993;178:1783–1788.

33 Stewart GA, Holt PG: Immunogenicity and tolerogenicity of a major house dust mite allergen, Der p 1 from *Dermatophagoides pteronyssinus*, in mice and rats. Int Arch Allergy Appl Immunol 1987;83:44–51.

34 Duez C, Tsicopoulos A, Janin A, Tillie-Leblond I, Thyphronitis G, Marquillies P, Hamid Q, Wallaert B, Tonnel AB, Pestel J: An in vivo model of allergic inflammation: pulmonary human cell infiltrate in allergen-challenged allergic hu-SCID mice. Eur J Immunol 1996;26:1088–1093.

35 Herz U, Schnoy N, Borelli S, Weigl L, Kasbohrer U, Daser A, Wahn U, Kottgen E, Renz H: A human-SCID mouse model for allergic immune response bacterial superantigen enhances skin inflammation and suppresses IgE production. J Invest Dermatol 1998;110:224–231.

36 Tournoy KG, Kips JC, Schou C, Pauwels RA: Airway eosinophilia is not a requirement for allergen-induced airway hyperresponsiveness. Clin Exp Allergy 2000;30:79–85.

37 Carvalho AF, Fusaro AE, Oliveira CR, Brito CA, Duarte AJ, Sato MN: *Blomia tropicalis* and *Dermatophagoides pteronyssinus* mites evoke distinct patterns of airway cellular influx in type I hypersensitivity murine model. J Clin Immunol 2004;24:533–541.

38 Gough L, Campbell E, Bayley D, Van Heeke G, Shakib F: Proteolytic activity of the house dust mite allergen Der p 1 enhances allergenicity in a mouse inhalation model. Clin Exp Allergy 2003;33:1159–1163.

39 Kikuchi Y, Takai T, Kuhara T, Ota M, Kato T, Hatanaka H, Ichikawa S, Tokura T, Akiba H, Mitsuishi K, Ikeda S, Okumura K, Ogawa H: Crucial commitment of proteolytic activity of a purified recombinant major house dust mite allergen Der p1 to sensitization toward IgE and IgG responses. J Immunol 2006;177:1609–1617.

40 Miller LA, Hurst SD, Coffman RL, Tyler NK, Stovall MY, Chou DL, Putney LF, Gershwin LJ, Schelegle ES, Plopper CG, Hyde DM: Airway generation-specific differences in the spatial distribution of immune cells and cytokines in allergen-challenged rhesus monkeys. Clin Exp Allergy 2005;35:894–906.

41 Lee YL, Ye YL, Yu CI, Wu YL, Lai YL, Ku PH, Hong RL, Chiang BL: Construction of single-chain interleukin-12 DNA plasmid to treat airway hyperresponsiveness in an animal model of asthma. Hum Gene Ther 2001;12:2065–2079.

42 Wilson MS, Taylor MD, Balic A, Finney CA, Lamb JR, Maizels RM: Suppression of allergic airway inflammation by helminth-induced regulatory T cells. J Exp Med 2005;202:1199–1212.

43 Yu CK, Shieh CM, Lei HY: Repeated intratracheal inoculation of house dust mite *(Dermatophagoides farinae)* induces pulmonary eosinophilic inflammation and IgE antibody production in mice. J Allergy Clin Immunol 1999;104:228–236.

44 Cates EC, Fattouh R, Wattie J, Inman MD, Goncharova S, Coyle AJ, Gutierrez-Ramos JC, Jordana M: Intranasal exposure of mice to house dust mite elicits allergic airway inflammation via a GM-CSF-mediated mechanism. J Immunol 2004;173:6384–6392.

45 King C, Brennan S, Thompson PJ, Stewart GA: Dust mite proteolytic allergens induce cytokine release from cultured airway epithelium. J Immunol 1998;161:3645–3651.

46 Gajewska BU, Wiley RE, Jordana M: GM-CSF and dendritic cells in allergic airway inflammation: basic mechanisms and prospects for therapeutic intervention. Curr Drug Targets 2003;2: 279–292.

47 Lewkowich IP, Herman NS, Schleifer KW, Dance MP, Chen BL, Dienger KM, Sproles AA, Shah JS, Kohl J, Belkaid Y, Wills-Karp M: CD4+CD25+ T cells protect against experimentally induced asthma and alter pulmonary dendritic cell phenotype and function. J Exp Med 2005;202: 1549–1561.

48 Huber HL, Koessler KK: The pathology of fatal asthma. Arch Intern Med 1922;30:689–760.

49 Zhen G, Park SW, Nguyenvu LT, Rodriguez MW, Barbeau R, Paquet AC, Erle DJ: Interleukin-13 and EGF receptor have critical but distinct roles in epithelial cell mucin production. Am J Respir Cell Mol Biol 2007;36:244–253.

50 Plante S, Semlali A, Joubert P, Bissonnette E, Laviolette M, Hamid Q, Chakir J: Mast cells regulate procollagen I (alpha 1) production by bronchial fibroblasts derived from subjects with asthma through IL-4/IL-4 delta 2 ratio. J Allergy Clin Immunol 2006;117:1321–1327.

51 Gomes I, Mathur SK, Espenshade BM, Mori Y, Varga J, Ackerman SJ: Eosinophil-fibroblast interactions induce fibroblast IL-6 secretion and extracellular matrix gene expression: implications in fibrogenesis. J Allergy Clin Immunol 2005;116:796–804.

52 Ramos-Barbon D, Presley JF, Hamid QA, Fixman ED, Martin JG: Antigen-specific CD4(+) T cells drive airway smooth muscle remodeling in experimental asthma. J Clin Invest 2005;115: 1580–1589.

53 Pegorier S, Wagner LA, Gleich GJ, Pretolani M: Eosinophil-derived cationic proteins activate the synthesis of remodeling factors by airway epithelial cells. J Immunol 2006;177:4861–4869.

54 Davies DE, Wicks J, Powell RM, Puddicombe SM, Holgate ST: Airway remodeling in asthma: new insights. J Allergy Clin Immunol 2003;111:215–226.

55 Holgate ST, Holloway J, Wilson S, Bucchieri F, Puddicombe S, Davies DE: Epithelial-mesenchymal communication in the pathogenesis of chronic asthma. Proc Am Thorac Soc 2004;1:93–98.

56 Temann UA, Geba GP, Rankin JA, Flavell RA: Expression of interleukin 9 in the lungs of transgenic mice causes airway inflammation, mast cell hyperplasia, and bronchial hyperresponsiveness. J Exp Med 1998;188:1307–1320.

57 Zhu Z, Homer RJ, Wang Z, Chen Q, Geba GP, Wang J, Zhang Y, Elias JA: Pulmonary expression of interleukin-13 causes inflammation, mucus hypersecretion, subepithelial fibrosis, physiologic abnormalities, and eotaxin production. J Clin Invest 1999;103:779–788.

58 Finotto S, Neurath MF, Glickman JN, Qin S, Lehr HA, Green FH, Ackerman K, Haley K, Galle PR, Szabo SJ, Drazen JM, De Sanctis GT, Glimcher LH: Development of spontaneous airway changes consistent with human asthma in mice lacking t-bet. Science 2002;295: 336–338.

59 Swirski FK, Sajic D, Robbins CS, Gajewska BU, Jordana M, Stämpfli MR: Chronic exposure to innocuous antigen in sensitized mice leads to suppressed airway eosinophilia that is reversed by granulocyte macrophage colony-stimulating factor. J Immunol 2002;169:3499–3506.

60 Henderson WR Jr, Tang LO, Chu SJ, Tsao SM, Chiang GK, Jones F, Jonas M, Pae C, Wang H, Chi EY: A role for cysteinyl leukotrienes in airway remodeling in a mouse asthma model. Am J Respir Crit Care Med 2002;165:108–116.

61 Tanaka H, Masuda T, Tokuoka S, Komai M, Nagao K, Takahashi Y, Nagai H: The effect of allergen-induced airway inflammation on airway remodeling in a murine model of allergic asthma. Inflamm Res 2001;50:616–624.

62 Temelkovski J, Hogan SP, Shepherd DP, Foster PS, Kumar RK: An improved murine model of asthma: selective airway inflammation, epithelial lesions and increased methacholine responsiveness following chronic exposure to aerosolised allergen. Thorax 1998;53:849–856.

63 Leigh R, Ellis R, Wattie J, Southam DS, De Hoogh M, Gauldie J, O'Byrne PM, Inman MD: Dysfunction and remodeling of the mouse airway persist after resolution of acute allergen-induced airway inflammation. Am J Respir Cell Mol Biol 2002;27:526–535.

64 Miller LA, Plopper CG, Hyde DM, Gerriets JE, Pieczarka EM, Tyler NK, Evans MJ, Gershwin LJ, Schelegle ES, Van Winkle LS: Immune and airway effects of house dust mite aeroallergen exposures during postnatal development of the infant rhesus monkey. Clin Exp Allergy 2003;33: 1686–1694.

65 Johnson JR, Wiley RE, Fattouh R, Swirski FK, Gajewska BU, Coyle AJ, Gutierrez-Ramos JC, Ellis R, Inman MD, Jordana M: Continuous exposure to house dust mite elicits chronic airway inflammation and structural remodeling. Am J Respir Crit Care Med 2004;169:378–385.

66 Angus AC, Ong ST, Chew FT: Sequence tag catalogs of dust mite-expressed genomes: utility in allergen and acarologic studies. Am J Pharmacogenomics 2004;4:357–369.

67 Krilis S, Baldo BA, Basten A: Antigens and allergens from the common house dust mite *Dermatophagoides pteronyssinus*. II. Identification of the major IgE-binding antigens by crossed radioimmunoelectrophoresis. J Allergy Clin Immunol 1984;74:142–146.

68 Lind P, Lowenstein H: Identification of allergens in *Dermatophagoides pteronyssinus* mite body extract by crossed radioimmunoelectrophoresis with two different rabbit antibody pools. Scand J Immunol 1983;17:263–273.

69 Tovey ER, Baldo BA: Comparison by electroblotting of IgE-binding components in extracts of house dust mite bodies and spent mite culture. J Allergy Clin Immunol 1987;79:93–102.

70 International Union of Immunologial Societies – Allergen Nomenclature Sub-Committee, 2006. http://www.Allergen.Org/.

71 Hewitt CR, Horton H, Jones RM, Pritchard DI: Heterogeneous proteolytic specificity and activity of the house dust mite proteinase allergen Der p 1. Clin Exp Allergy 1997;27:201–207.

72 King C, Simpson RJ, Moritz RL, Reed GE, Thompson PJ, Stewart GA: The isolation and characterization of a novel collagenolytic serine protease allergen (Der p 9) from the dust mite *Dermatophagoides pteronyssinus*. J Allergy Clin Immunol 1996;98:739–747.

73 Stewart GA, Ward LD, Simpson RJ, Thompson PJ: The group III allergen from the house dust mite *Dermatophagoides pteronyssinus* is a trypsin-like enzyme. Immunology 1992;75:29–35.

74 Yasueda H, Mita H, Akiyama K, Shida T, Ando T, Sugiyama S, Yamakawa H: Allergens from *Dermatophagoides* mites with chymotryptic activity. Clin Exp Allergy 1993;23:384–390.

75 Wan H, Winton HL, Soeller C, Gruenert DC, Thompson PJ, Cannell MB, Stewart GA, Garrod DR, Robinson C: Quantitative structural and biochemical analyses of tight junction dynamics following exposure of epithelial cells to house dust mite allergen Der p 1. Clin Exp Allergy 2000;30: 685–698.

76 Wan H, Winton HL, Soeller C, Tovey ER, Gruenert DC, Thompson PJ, Stewart GA, Taylor GW, Garrod DR, Cannell MB, Robinson C: Der p 1 facilitates transepithelial allergen delivery by disruption of tight junctions. J Clin Invest 1999;104:123–133.

77 Wan H, Winton HL, Soeller C, Taylor GW, Gruenert DC, Thompson PJ, Cannell MB, Stewart GA, Garrod DR, Robinson C: The transmembrane protein occludin of epithelial tight junctions is a functional target for serine peptidases from faecal pellets of *Dermatophagoides pteronyssinus*. Clin Exp Allergy 2001;31:279–294.

78 Chen CL, Lee CT, Liu YC, Wang JY, Lei HY, Yu CK: House dust mite *Dermatophagoides farinae* augments proinflammatory mediator productions and accessory function of alveolar macrophages: implications for allergic sensitization and inflammation. J Immunol 2003;170: 528–536.

79 Sun G, Stacey MA, Schmidt M, Mori L, Mattoli S: Interaction of mite allergens Der p 3 and Der p 9 with protease-activated receptor-2 expressed by lung epithelial cells. J Immunol 2001;167: 1014–1021.

80 Asokananthan N, Graham PT, Stewart DJ, Bakker AJ, Eidne KA, Thompson PJ, Stewart GA: House dust mite allergens induce proinflammatory cytokines from respiratory epithelial cells: the cysteine protease allergen, Der p 1, activates protease-activated receptor (PAR)-2 and inactivates PAR-1. J Immunol 2002;169:4572–4578.

81 Adam E, Hansen KK, Astudillo OF, Coulon L, Bex F, Duhant X, Jaumotte E, Hollenberg MD, Jacquet A: The house dust mite allergen Der p 1, unlike Der p 3, stimulates the expression of interleukin-8 in human airway epithelial cells via a proteinase-activated receptor-2-independent mechanism. J Biol Chem 2006;281:6910–6923.

82 Kauffman HF, Tamm M, Timmerman JA, Borger P: House dust mite major allergens Der p 1 and Der p 5 activate human airway-derived epithelial cells by protease-dependent and protease-independent mechanisms. Clin Mol Allergy 2006;4:5.

83 Hewitt CR, Brown AP, Hart BJ, Pritchard DI: A major house dust mite allergen disrupts the immunoglobulin E network by selectively cleaving CD23: innate protection by antiproteases. J Exp Med 1995;182:1537–1544.

84 Schulz O, Sutton BJ, Beavil RL, Shi J, Sewell HF, Gould HJ, Laing P, Shakib F: Cleavage of the low-affinity receptor for human IgE (CD23) by a mite cysteine protease: nature of the cleaved fragment in relation to the structure and function of CD23. Eur J Immunol 1997;27:584–588.

85 Luo HY, Hofstetter H, Banchereau J, Delespesse G: Cross-linking of CD23 antigen by its natural ligand (IgE) or by anti-CD23 antibody prevents B lymphocyte proliferation and differentiation. J Immunol 1991;146:2122–2129.

86 Yu P, Kosco-Vilbois M, Richards M, Kohler G, Lamers MC: Negative feedback regulation of IgE synthesis by murine CD23. Nature 1994;369:753–756.

87 Pene J, Chretien I, Rousset F, Briere F, Bonnefoy JY, de Vries JE: Modulation of IL-4-induced human IgE production in vitro by IFN-gamma and IL-5: the role of soluble CD23 (s-CD23). J Cell Biochem 1989;39:253–264.

88 Sarfati M, Delespesse G: Possible role of human lymphocyte receptor for IgE (CD23) or its soluble fragments in the in vitro synthesis of human IgE. J Immunol 1988;141:2195–2199.

89 Saxon A, Ke Z, Bahati L, Stevens RH: Soluble CD23 containing B cell supernatants induce IgE from peripheral blood B-lymphocytes and costimulate with interleukin-4 in induction of IgE. J Allergy Clin Immunol 1990;86:333–344.

90   Machado DC, Horton D, Harrop R, Peachell PT, Helm BA: Potential allergens stimulate the release of mediators of the allergic response from cells of mast cell lineage in the absence of sensitization with antigen-specific IgE. Eur J Immunol 1996;26:2972–2980.

91   Schulz O, Sewell HF, Shakib F: Proteolytic cleavage of CD25, the alpha subunit of the human T cell interleukin 2 receptor, by Der p 1, a major mite allergen with cysteine protease activity. J Exp Med 1998;187:271–275.

92   Ghaemmaghami AM, Robins A, Gough L, Sewell HF, Shakib F: Human T cell subset commitment determined by the intrinsic property of antigen: the proteolytic activity of the major mite allergen Der p 1 conditions T cells to produce more IL-4 and less IFN-gamma. Eur J Immunol 2001;31: 1211–1216.

93   Gough L, Schulz O, Sewell HF, Shakib F: The cysteine protease activity of the major dust mite allergen Der p 1 selectively enhances the immunoglobulin E antibody response. J Exp Med 1999;190:1897–1902.

94   Zhu Z, Zheng T, Homer RJ, Kim YK, Chen NY, Cohn L, Hamid Q, Elias JA: Acidic mammalian chitinase in asthmatic Th2 inflammation and IL-13 pathway activation. Science 2004;304: 1678–1682.

95   Bleau G, Massicotte F, Merlen Y, Boisvert C: Mammalian chitinase-like proteins. EXS 1999;87: 211–221.

96   Boot RG, Blommaart EF, Swart E, Ghauharali-van der Vlugt K, Bijl N, Moe C, Place A, Aerts JM: Identification of a novel acidic mammalian chitinase distinct from chitotriosidase. J Biol Chem 2001;276:6770–6778.

97   Boot RG, Renkema GH, Verhoek M, Strijland A, Bliek J, de Meulemeester TM, Mannens MM, Aerts JM: The human chitotriosidase gene. Nature of inherited enzyme deficiency. J Biol Chem 1998;273:25680–25685.

98   Elias JA, Homer RJ, Hamid Q, Lee CG: Chitinases and chitinase-like proteins in T(h)2 inflammation and asthma. J Allergy Clin Immunol 2005;116:497–500.

99   Chang NC, Hung SI, Hwa KY, Kato I, Chen JE, Liu CH, Chang AC: A macrophage protein, ym1, transiently expressed during inflammation is a novel mammalian lectin. J Biol Chem 2001;276: 17497–17506.

100 Owhashi M, Arita H, Hayai N: Identification of a novel eosinophil chemotactic cytokine (ECF-L) as a chitinase family protein. J Biol Chem 2000;275:1279–1286.

101 McCall C, Hunter S, Stedman K, Weber E, Hillier A, Bozic C, Rivoire B, Olivry T: Characterization and cloning of a major high molecular weight house dust mite allergen (Der f 15) for dogs. Vet Immunol Immunopathol 2001;78:231–247.

102 Weber E, Hunter S, Stedman K, Dreitz S, Olivry T, Hillier A, McCall C: Identification, characterization, and cloning of a complementary DNA encoding a 60-kd house dust mite allergen (Der f 18) for human beings and dogs. J Allergy Clin Immunol 2003;112:79–86.

103 Noli C, Bernadina WE, Willemse T: The significance of reactions to purified fractions of *Dermatophagoides pteronyssinus* and *Dermatophagoides farinae* in canine atopic dermatitis. Vet Immunol Immunopathol 1996;52:147–157.

104 O'Neil SE, Heinrich TK, Hales BJ, Hazell LA, Holt DC, Fischer K, Thomas WR: The chitinase allergens Der p 15 and Der p 18 from *Dermatophagoides pteronyssinus*. Clin Exp Allergy 2006;36:831–839.

105 Aki T, Kodama T, Fujikawa A, Miura K, Shigeta S, Wada T, Jyo T, Murooka Y, Oka S, Ono K: Immunochemical characterization of recombinant and native tropomyosins as a new allergen from the house dust mite, *Dermatophagoides farinae*. J Allergy Clin Immunol 1995;96:74–83.

106 Tsai LC, Chao PL, Shen HD, Tang RB, Chang TC, Chang ZN, Hung MW, Lee BL, Chua KY: Isolation and characterization of a novel 98-kd *Dermatophagoides farinae* mite allergen. J Allergy Clin Immunol 1998;102:295–303.

107 Epton MJ, Dilworth RJ, Smith W, Hart BJ, Thomas WR: High-molecular-weight allergens of the house dust mite: an apolipophorin-like cDNA has sequence identity with the major M-177 allergen and the IgE-binding peptide fragments Mag1 and Mag3. Int Arch Allergy Immunol 1999;120:185–191.

108 Fujikawa A, Ishimaru N, Seto A, Yamada H, Aki T, Shigeta S, Wada T, Jyo T, Murooka Y, Oka S, Ono K: Cloning and characterization of a new allergen, Mag 3, from the house dust mite,

Dermatophagoides farinae: cross-reactivity with high-molecular-weight allergen. Mol Immunol 1996;33:311–319.

109 Lapa e Silva JR, Possebon da Silva MD, Lefort J, Vargaftig BB: Endotoxins, asthma, and allergic immune responses. Toxicology 2000;152:31–35.

110 Reed CE, Milton DK: Endotoxin-stimulated innate immunity: a contributing factor for asthma. J Allergy Clin Immunol 2001;108:157–166.

111 Ritz SA, Stämpfli MR, Davies DE, Holgate ST, Jordana M: On the generation of allergic airway diseases: from GM-CSF to Kyoto. Trends Immunol 2002;23:396–402.

112 Eisenbarth SC, Piggott DA, Huleatt JW, Visintin I, Herrick CA, Bottomly K: Lipopolysaccharide-enhanced, toll-like receptor 4-dependent T helper cell type 2 responses to inhaled antigen. J Exp Med 2002;196:1645–1651.

113 Tulic MK, Wale JL, Holt PG, Sly PD: Modification of the inflammatory response to allergen challenge after exposure to bacterial lipopolysaccharide. Am J Respir Cell Mol Biol 2000;22:604–612.

114 Chisholm D, Libet L, Hayashi T, Horner AA: Airway peptidoglycan and immunostimulatory DNA exposures have divergent effects on the development of airway allergen hypersensitivities. J Allergy Clin Immunol 2004;113:448–454.

115 McMillan SJ, Xanthou G, Lloyd CM: Manipulation of allergen-induced airway remodeling by treatment with anti-TGF-beta antibody: effect on the SMAD signaling pathway. J Immunol 2005;174:5774–5780.

116 Fattouh R, Midence NG, Arias K, Johnson JR, Lonning S, Gauldie J, Jordana M: Impact of TGF-beta blockade on house dust mite-induced allergic airways inflammation on remodeling (abstract). Meeting of the American Thoracic Society, San Francisco, 2007.

117 Tovey ER, Mitakakis TZ, Sercombe JK, Vanlaar CH, Marks GB: Four methods of sampling for dust mite allergen: differences in 'dust'. Allergy 2003;58:790–794.

118 Custis NJ, Woodfolk JA, Vaughan JW, Platts-Mills TA: Quantitative measurement of airborne allergens from dust mites, dogs, and cats using an ion-charging device. Clin Exp Allergy 2003;33:986–991.

119 Platts-Mills TA, Vervloet D, Thomas WR, Aalberse RC, Chapman MD: Indoor allergens and asthma: report of the third international workshop. J Allergy Clin Immunol 1997;100:S2–S24.

120 Sakaguchi M, Inouye S, Sasaki R, Hashimoto M, Kobayashi C, Yasueda H: Measurement of airborne mite allergen exposure in individual subjects. J Allergy Clin Immunol 1996;97:1040–1044.

121 Colloff MJ: Distribution and abundance of dust mites within homes. Allergy 1998;53:24–27.

122 Sakaguchi M, Inouye S, Yasueda H, Shida T: Concentration of airborne mite allergens (Der I and Der II) during sleep. Allergy 1992;47:55–57.

123 Sakaguchi M, Inouye S, Yasueda H, Irie T, Yoshizawa S, Shida T: Measurement of allergens associated with dust mite allergy. II. Concentrations of airborne mite allergens (Der I and Der II) in the house. Int Arch Allergy Appl Immunol 1989;90:190–193.

124 Zock JP, Heinrich J, Jarvis D, Verlato G, Norback D, Plana E, Sunyer J, Chinn S, Olivieri M, Soon A, Villani S, Ponzio M, Dahlman-Hoglund A, Svanes C, Luczynska C: Distribution and determinants of house dust mite allergens in Europe: the European community respiratory health survey II. J Allergy Clin Immunol 2006;118:682–690.

125 Lau S, Illi S, Sommerfeld C, Niggemann B, Bergmann R, von Mutius E, Wahn U: Early exposure to house-dust mite and cat allergens and development of childhood asthma: a cohort study. Multicentre Allergy Study Group. Lancet 2000;356:1392–1397.

126 Sporik R, Holgate ST, Platts-Mills TA, Cogswell JJ: Exposure to house-dust mite allergen (Der p 1) and the development of asthma in childhood. A prospective study. N Engl J Med 1990;323:502–507.

127 Owen S, Morganstern M, Hepworth J, Woodcock A: Control of house dust mite antigen in bedding. Lancet 1990;335:396–397.

128 Leaderer BP, Belanger K, Triche E, Holford T, Gold DR, Kim Y, Jankun T, Ren P, McSharry, Je JE, Platts-Mills TA, Chapman MD, Bracken MB: Dust mite, cockroach, cat, and dog allergen concentrations in homes of asthmatic children in the northeastern United States: impact of socioeconomic factors and population density. Environ Health Perspect 2002;110:419–425.

129 Asher MI, Keil U, Anderson HR, Beasley R, Crane J, Martinez F, Mitchell EA, Pearce N, Sibbald B, Stewart AW, et al: International Study of Asthma and Allergies in Childhood (ISAAC): rationale and methods. Eur Respir J 1995;8:483–491.

130 Peat JK, Tovey E, Toelle BG, Haby MM, Gray EJ, Mahmic A, Woolcock AJ: House dust mite allergens. A major risk factor for childhood asthma in Australia. Am J Respir Crit Care Med 1996;153: 141–146.

131 Wickens K, Siebers R, Ellis I, Lewis S, Sawyer G, Tohill S, Stone L, Kent R, Kennedy J, Slater T, Crothall A, Trethowen H, Pearce N, Fitzharris P, Crane J: Determinants of house dust mite allergen in homes in Wellington, New Zealand. Clin Exp Allergy 1997;27:1077–1085.

132 Bemt L, Vries MP, Knapen L, Jansen M, Goossens M, Muris JW, Schayck CP: Influence of mattress characteristics on house dust mite allergen concentration. Clin Exp Allergy 2006;36: 233–237.

133 Torrent M, Sunyer J, Munoz L, Cullinan P, Iturriaga MV, Figueroa C, Vall O, Taylor AN, Anto JM: Early-life domestic aeroallergen exposure and IgE sensitization at age 4 years. J Allergy Clin Immunol 2006;118:742–748.

134 van Strien RT, Koopman LP, Kerkhof M, Spithoven J, de Jongste JC, Gerritsen J, Neijens HJ, Aalberse RC, Smit HA, Brunekreef B: Mite and pet allergen levels in homes of children born to allergic and nonallergic parents: The PIAMA study. Environ Health Perspect 2002;110: A693–A698.

135 Southam DS, Dolovich M, O'Byrne PM, Inman MD: Distribution of intranasal instillations in mice: effects of volume, time, body position, and anesthesia. Am J Physiol Lung Cell Mol Physiol 2002;282:L833–L839.

136 Platts-Mills TA, Sporik RB, Wheatley LM, Heymann PW: Is there a dose-response relationship between exposure to indoor allergens and symptoms of asthma? J Allergy Clin Immunol 1995;96: 435–440.

137 Sporik R, Platts-Mills TA: Allergen exposure and the development of asthma. Thorax 2001;56 (suppl 2):ii58–63.

138 Upham JW, Holt PG: Environment and development of atopy. Curr Opin Allergy Clin Immunol 2005;5:167–172.

139 Kuehr J, Frischer T, Meinert R, Barth R, Forster J, Schraub S, Urbanek R, Karmaus W: Mite allergen exposure is a risk for the incidence of specific sensitization. J Allergy Clin Immunol 1994;94:44–52.

Manel Jordana, MD, PhD
MDCL 4013, Department of Pathology and Molecular Medicine, McMaster University
1200 Main Street West
Hamilton, ON L8S 3Z5 (Canada)
Tel. +1 905 525 9140, ext. 22473, Fax +1 905 522 6750, E-Mail jordanam@mcmaster.ca

Sjöbring U, Taylor JD (eds): Models of Exacerbations in Asthma and COPD.
Contrib Microbiol. Basel, Karger, 2007, vol 14, pp 68–82

Respiratory Syncytial Virus-Induced Pulmonary Disease and Exacerbation of Allergic Asthma

Nicholas W. Lukacs, Joost Smit, Dennis Lindell, Matthew Schaller

Department of Pathology, University of Michigan Medical School,
Ann Arbor, Mich., USA

Abstract

Several respiratory viruses have been shown to cause exacerbations of asthma. While the various viral responses likely have common mechanisms of activation, the respiratory syncytial virus (RSV) appears to promote specific responses that on their own can cause severe pulmonary problems. Understanding the mechanisms that promote inappropriate immune responses and local damage may lead to better therapy. The activation and recruitment of T cells that amplify and skew the immune response toward more intense pathology, including mucus production and remodeling of the airways, are likely scenarios that lead to more severe disease and clinical crisis in asthmatic patients. These mechanisms may also contribute to a significant proportion of exacerbations in chronic obstructive pulmonary disease. This review will focus on recent research on specific pathways of RSV-mediated activation of the innate host defense, including chemokine biology and TLR pathways, as well as on acquired immunity.

Introduction

Several epidemiological studies have shown that respiratory viral infection is an independent risk factor for the development of severe chronic asthma exacerbations [1–5]. Although a number of respiratory viruses have been implicated in the induction of asthma exacerbations, including infections with rhinovirus, influenza, parainfluenza, and adenovirus, recent studies in children [6] and adults [7] with severe asthmatic exacerbations (for definitions see the introductory chapter of this book) often are associated with respiratory syncytial virus (RSV) infections (27 and 37%, respectively in the two studies). In particular, both of these studies indicated that in the most severe patients primarily two viral infections were involved, RSV or influenza A, with RSV being the most commonly associated infection. In addition to the role in asthma, this virus poses a major threat to young children as the major proportion of infections that hospitalize infants (~90%) are caused by RSV infections [8, 9].

RSV epidemics occur annually in temperate climates, such as in the US, where two thirds of infants are infected within their first year of life with ~2% hospitalized. It appears that individuals can be reinfected multiple times with the same strain. RSV infections are now recognized as the No. 1 cause of childhood hospitalization in the US with the most vulnerable infants having significant length of stay in the ICU, further driving up health costs. It is estimated by the CDC that up to 125,000 pediatric hospitalizations in the United States each year are due to RSV, at an annual cost of over USD 300,000,000 [10]. Despite the generation of RSV-specific adaptive immune responses, RSV does not confer protective immunity and recurrent infections throughout life are common [11, 12]. While RSV is especially detrimental in very young infants whose airways are small and easily occluded, causing bronchiolitis or pneumonitis, RSV is also widely becoming recognized as an important pathogen in lung transplant recipients, with exacerbations in patients with chronic obstructive pulmonary disease, as well as other patients with chronic lung disease, especially asthma and in the elderly population in general. Combining the figures for mortality for all ages results in approximately 30/100,000 from 1990 to 2000, with an annual average of over 17,000 in the US [13, 14]. These numbers are likely grossly underestimated, as RSV infection is not diagnosed in adults in a consistent manner. Thus, RSV not only causes significant exacerbated lung disease in young and old, but also is associated with a significant amount of mortality directly.

Although anti-RSV antibodies are available and appear to alleviate severe disease, they perform best when given prophylactically. Few other options exist for combating the RSV infections in susceptible patient populations [15–20]. In the late 1960s, attempts to vaccinate children with an alum-precipitated formalin-inactivated RSV vaccine preparation failed and caused severe exacerbated disease upon reinfection with live RSV. The clinical manifestations appeared to be a result of an enhanced Th2 disease, mucus production and eosinophilia, not observed in nonvaccinated children.

Respiratory Syncytial Virus

RSV is an enveloped, nonsegmented negative-sense single-stranded RNA (ssRNA) virus. The RSV genome is composed of ~15,000 nucleotides that are transcribed into 11 different viral proteins. Vaccine development against recombinant surface proteins (G, F, and SH) has failed and when whole virus was used inoculation actually led to exacerbation of severe pulmonary disease upon RSV infection. In animal models the adverse responses induced by the recombinant vaccines or fixed virus were characterized by an exaggerated Th2 response eosinophilia and mucus overproduction [21–23]. Thus, the proteins from RSV, as well as the virus itself, can induce Th2-type responses that may dictate the progression into chronic disease. Epidemiological studies support the concept that the detrimental responses observed in infants and young children are associated with the development of childhood asthma (and may be a determinant of susceptibility to exacerbations of asthmatic disease). One consequence of an early severe RSV infection appears to be long-term pulmonary problems even years after the infection has resolved [9, 13, 24]. It is unclear, however, whether these long-term problems are caused, intensified or merely uncovered by the RSV infection. It may be that RSV helps to establish an immune environment in the lung that allows a progression into chronic pulmonary problems in immunologically immature, genetically susceptible individuals. Importantly, it is these same immune responses that are induced by RSV that likely lead to the most severe exacerbations in asthmatic individuals.

Innate Immune Response to RSV Infection

To progress with studies of RSV-induced asthma exacerbation, it is important to understand the immune responses to RSV itself. In children with severe RSV infection a number of leukocyte populations, including neutrophils, macrophages, lymphocytes, and eosinophils, can be identified from airway samples [3, 25]. Using animal models of disease to examine the temporal expression of leukocyte accumulation, the cellular response within the first 4 days appears to consist of macrophages, neutrophils, and NK cells [26, 27]. The differential activation of this initial cell response may be critical to establish an effective antiviral immune reaction. The subsequent migration and activation of the adaptive immune cells, such cytotoxic CD8+ T cells, likely depend upon the 'proper' activation of these cell populations [28–31]. Previous studies have demonstrated that the CD8+ T cell population in BALB/c mice is defective and does not properly clear virally infected cells [28, 32, 33]. When these latter immune events do not lead to an antiviral environment, the system may attempt to inhibit the virus using other mechanisms.

One of the most important pathological mechanisms in severely infected infants during RSV infection is the dysregulation of mucus production within the airway. Mucus production can be beneficial in an innate response as it can function to inhibit reinfection of the virus as well as provide a protective environment for the epithelial cell layer [34]. However, if overproduced by inappropriate responses such as by IL-13-driven mechanisms, mucus can inhibit gas exchange in the lung. The overproduction of mucus is especially detrimental in infants whose airways are small and can be easily obstructed [9, 35–37]. These pathological reactions to the RSV are seen in acute bronchiolitis and are also likely to be involved, at least in part, in asthmatic exacerbation.

The decision to respond properly to the RSV infection must be made very early (i.e. prior to virion production) in order to limit the viral propagation. The initiation of an effective resistance to viruses is dependent upon the early innate immune response to recognize the presence of foreign pathogens. The innate immune response is the first line of defense and is responsible for recognizing the vast array of pathogens with a limited set of fixed germline-encoded receptors. To deal with this challenge, innate immunity relies on the detection of pathogen-associated molecular patterns unique to various classes of pathogens. Bacteria and fungi possess specific invariant structures like lipopolysaccharide and zymosan that allow for easier discrimination of 'foreignness' than viruses [38, 39]. In contrast, the heterogeneity of viral glycoproteins, along with their ability to drift genetically from season to season, creates even greater challenges for innate immune recognition of viruses. To circumvent these obstacles, the innate immune system has evolved mechanisms to detect characteristics of viral nucleic acids that are either distinct in structure (double-stranded RNA, dsRNA) or subcellular location (ssRNA). Since RSV is an ssRNA negative sense virus, both ssRNA and dsRNA species are formed and are targets for the innate immune system (fig. 1). A subset of pattern recognition receptors includes the toll-like receptors (TLR), which recognize different pathogen-associated molecular patterns and activate NF-κB and other innate signaling pathways [40–42]. Previous studies have suggested that RSV may bind to TLR4 to initiate the antiviral responses [43, 44]. Recent clinical data has suggested that distinct polymorphisms in TLR4 genes in humans can be linked to the susceptibility to severe RSV infection and also to long-term consequences in pulmonary function [45]. However, in order to transcribe and translate the protein components of RSV for assembly, the virus must go through a dsRNA step. It is recognition of this stage by another TLR, TLR3, that likely leads to the recognition of the infection by the host immune response. The activation of TLR3 in dendritic cells (DC) using poly-IC (an ssRNA mimetic) leads to production of IL-12 and type I IFN that are required for initiation of a cytotoxic IFN-γ-mediated response [46, 47]. Recent studies from our lab using TLR3–/– mice have established that RSV infection creates an altered immune environment that results in the upregulation of IL-13 and overproduction of mucus [48]. Other innate molecules can also be implicated in the initiation of this response, including the two

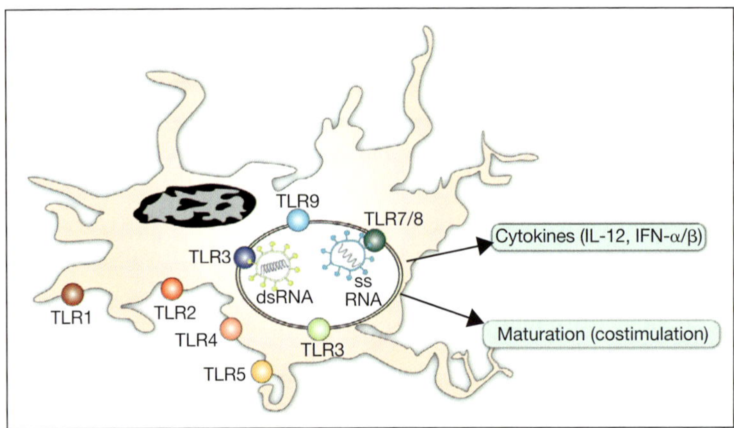

Fig. 1. DC recognize dsRNA (TLR3) and ssRNA (TLR7) during RSV infection and promote antiviral responses.

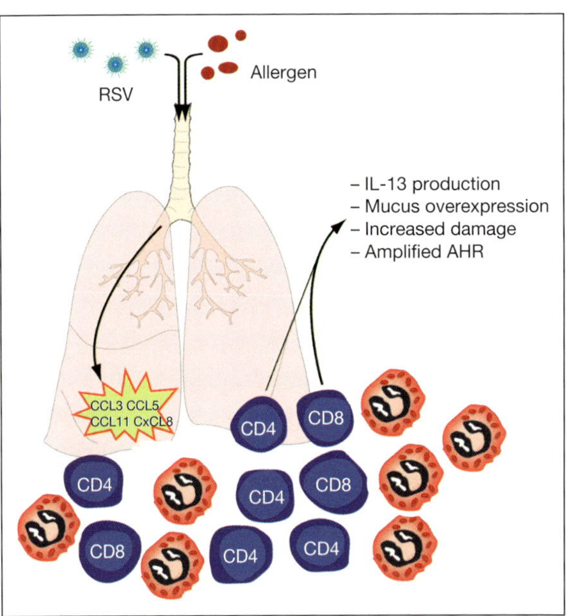

Fig. 2. Exacerbation of asthma by RSV induced chemokine production and recruitment of lymphocytes and eosinophils. AHR = Airway hyperresponsiveness.

receptors TLR7/TLR8 that recognize ssRNA as well as RIG-I, a cytoplasmic receptor for dsRNA [49–52]. Together these molecules can mediate the proper antiviral response by inducing type I IFNs and IL-12. Clinical studies have suggested that one critical aspect that correlates to effective viral response was the ability to produce sufficient IL-12 from immune cells [53]. IL-12 not only mediates IFN-γ-associated responses, but can directly be involved in effective antiviral immunity [54–56]. In addition, these type 1 immune responses have been shown to diminish asthma responses by altering the local pulmonary immune environment. It may be that these important innate immune mechanisms are inappropriately regulated in asthma due to ongoing Th2 responses.

Role of DC Subsets in RSV-Induced Exacerbation

DCs are key antigen-presenting cells that link the innate and adaptive immune systems. Early studies demonstrated that increased numbers of DCs are recruited to the airway of asthmatics during severe disease and that corticosteroid treatment could control their recruitment [57, 58]. Furthermore, studies in animal models indicated that DCs were required for the development of chronic eosinophil inflammation in the airway [59]. Two major DCs subsets are the CD11b+, CD11c+ myeloid/conventional DC (mDC/cDC) and CD11b–, B220+ plasmacytoid DC (pDC) [60]. A series of studies have outlined distinct differences in the ability of DC subsets to induce responses. In particular, cDC have been implicated in driving a proallergic response [61], while pDC have been identified to block or 'tolerize' the pulmonary immune environment against Th2 responses [62]. Other reports have demonstrated that pDC can participate in or even functionally mediate antiviral responses in the lung, either directly or indirectly through activation of other cell populations [63–66]. Importantly, a number of studies have now shown that RSV infection of DCs significantly alters the ability of mDC to express class II molecules, express costimulatory molecules, preferentially express IL-10, and impair induction of a Th1 response [67–71]. Thus, in an ongoing RSV infection, antigens (allergens) may induce a skewed Th2 response due to the alteration of DC subset activation. Results from our own laboratory have indicated that depletion of pDC from the lungs of RSV-infected mice results in an enhanced pathological response with increased airway hyperresponsiveness, mucus, and skewing toward a Th2 cytokine profile [72]. Thus, the decision to recruit the proper DC subset into the lung during a response may be important for controlling the severity of RSV infection. Studies have clearly established that both pDC and mDC subsets are recruited to the airway during RSV infection and the severity of disease may depend upon the numbers of each subset that respond to the stimulus along with the level of exposure to allergen [70, 71, 73]. Chemokine receptors on DCs play important roles for localization and positioning the cells in the proper alignment for monitoring or responding to infectious and noxious stimuli. In the lung, it is not clear what receptors are involved in each phase of the process but recent studies have suggested that CxCR3 has an important role in the movement of pDC from the lung to the lymph node during a respiratory viral infection [74]. Thus, activation of particular patterns of chemokines may help set up an environment for recruitment of the correct innate immune cell populations and therefore influence the direction of the immune response.

Chemokines and RSV Infection

Clinical studies have suggested that the severity of RSV-induced disease correlates with the influx of leukocytes [11]. Chemokines (chemotactic cytokines) have been shown to correlate directly to the intensity of the inflammatory response and can be induced by viral infection in multiple pulmonary cell populations. In particular, a number of studies demonstrated that the levels of CxCL8 (IL-8), CCL3, and CCL5 produced during the RSV infection correlates to the severity of RSV infection in infants [75–77]. Local alveolar macrophage populations have demonstrated the ability to produce a number of activating and inflammatory cytokines, such as TNF, IL-1, IL-6, and IL-8, in response to RSV infection in vitro [78–81]. These cytokines may subsequently activate other cells within the airway and amplify the activation responses via cytokine networks leading to inflammation-associated damage within the airway. RSV-infected epithelial cells are a significant source of a number of chemokines, including CCL2, CCL3, CCL5, and CxCL8 [82–84], the expression of which are positively regulated by NF-κB [85]. Thus, macrophages and epithelial cells directly, through infection with RSV, induce an amplified system of activating and chemoattractant cytokines. It is not known whether individuals who are susceptible to severe RSV infections, versus those that are not, have different profiles of cytokines and chemokines. However, in infants with RSV infection high levels of chemokines relate to increased severity of the disease [86, 87]. Several of these chemokines have been associated with inflammatory responses in animal models of RSV infection. Domachowske et al. [88] demonstrated that infection of mice with paramyxovirus, a virus in the same family as RSV, induces increased CCL3 which in turn was related to granulocyte recruitment. CCL5, another chemokine, has a significant impact on the pathophysiological response both in primary RSV infection and during exacerbation of allergic airway inflammation enhanced by RSV [89, 90]. In previous studies we have also found that the overproduction of mucus and development of airway hyperreactivity is directly related to the expression and activation of CxCR2 (an IL-8 receptor homologue) in mice [91]. In addition, an interferon-inducible CxC chemokine, IP-10 (CxCL10), is also induced during RSV infection [92–94] and has been identified by our lab and others to be associated with allergic pulmonary disease [95–97]. The production of these chemokines during RSV infection may be predominantly due to innate molecule-mediated activation via TLR-associated responses [98–100]. Thus, overproduction of chemokines is likely to contribute strongly to the severity of RSV infection and may also impact on cellular recruitment during exacerbations in allergic responses.

Asthma, Virus Infection, and Recruitment of T Cells

The increased incidence in asthma over recent years has been linked to multiple factors [101, 102]. However, a disproportionate increase has been observed in inner city children. Recent studies have identified that many of the severe asthmatic cases in the inner city can be linked to cockroach allergen sensitivity [103–105]. Similar to the house dust mite allergen (reviewed in a previous chapter), the cockroach-derived antigens appear to induce allergic responses that lead to airway inflammation and subsequently to changes in airway physiology. There is substantial evidence that these allergic inflammatory responses are driven by Th2-type cytokines including IL-4, IL-5 and IL-13 [106–108]. The mechanisms governing the preferential recruitment and activation of these Th2 cells have been a primary focus directed at discovering therapeutic strategies. Recent data have indicated that in the most severe asthmatics there is an alteration in the ratio of CD4:CD8 T cells found in the airway, such that the ratio is significantly reduced compared to either nonasthmatics or mild asthmatics with an increase in CD8+ T cells [109, 110]. In addition, a number of studies have identified that CD8+ T cells play a role for the response both to allergens and to viruses in animal models [111–114]. Thus, the role of CD8 T cells for asthma exacerbation has become an area of greater interest due to the realization that these cells may influence Th2-type cytokines during the response of asthmatic individuals to virus infection [110, 115, 116]. Understanding whether these individual T cell subsets observed in viral exacerbations of asthma are specifically contributing to disease will be important for identifying novel targets for treatment. The recruitment of T cell subsets to the airways appears to be differentially regulated by chemokines produced during inflammatory responses.

The identification of chemokines in the airways of asthmatics after allergen provocation indicated that they play a significant role in the accumulation of leukocytes [117–119]. A wide array of data within the allergic asthma field has further defined potential chemokines and their receptors using animal models. Most of these analyses have focused on chemokine receptors on CD4+ T cells. Almost no information has defined the receptor expression on CD8 cells and their role in asthmatic inflammation. Interestingly, there is substantial information available that indicates that there may be preferential chemokine utilization that relates to allergic/asthmatic responses. Along with Th2 cytokine-driven chemokines, there may be preferential expression of chemokine receptors on effector cells that migrate into the airway that can induce damage leading to airway hyperreactivity. In particular, studies have demonstrated a link between IL-13-induced remodeling and CCR1 in fungal challenge [120] and using IL-13 transgenic mice crossed onto a CCR1-deficient background [121]. Recent data from our laboratory demonstrate that CCR1+ CD8+ T cells are recruited only after RSV-induced asthma exacerbation and appear to provide an important activation role that results in increased levels of Th2 cytokines (unpubl. data). The RSV-induced stimulation appears to provide a mechanism that drives increased T cell recruitment through the production of high levels of CCR1-binding chemokines, especially CCL5. Other chemokine receptors, including CCR6 and CxCR2, may also play important roles. Thus, exacerbation of asthma by RSV may be a result of inducing chemokines that recruit inappropriately responding cell populations that exacerbate the asthma-like inflammation.

Closing Remarks

While the mechanisms and particular viruses that mediate exacerbation of asthma have not been completely defined, it is likely that certain pathways are common to multiple respiratory viral infections. These pathways likely include the initial activation of chemokine production via innate molecule recognition systems, such as TLRs, followed by intensification of an already existing chronic inflammatory process. Subsequently, the activation and recruitment of T cells that amplify and skew the immune response toward more intense pathology, including mucus production and remodeling of the airways, are likely scenarios that lead to more severe disease and clinical crisis. RSV has a unique ability to influence all of these pathways and may therefore present a particularly difficult challenge. A better understanding of this process will allow a long-term strategy for targeting specific mediators or receptors that could alleviate progression of acute disease episodes and chronic disease phenotypes mediated by RSV.

References

1   Busse WW, Gern JE, Dick EC: The role of respiratory viruses in asthma. Ciba Found Symp 1997;206:208–219.

2   Stempel DA, Boucher RC: Respiratory infection and airway reactivity. Med Clin North Am 1981;65:1045–1053.

3   Hogg JC: Childhood viral infection and the pathogenesis of asthma and chronic obstructive lung disease. Am J Respir Crit Care Med 1999;160:S26–S28.

4   Holtzman MJ, Morton JD, Shornick LP, Tyner JW, O'Sullivan MP, Antao A, Lo M, Castro M, Walter MJ: Immunity, inflammation, and remodeling in the airway epithelial barrier: epithelial-viral-allergic paradigm. Physiol Rev 2002;82:19–46.

5   Nicholson KG, Kent J, Ireland DC: Respiratory viruses and exacerbations of asthma in adults. BMJ 1993;307:982–986.

6   Zhao J, Takamura M, Yamaoka A, Odajima Y, Iikura Y: Altered eosinophil levels as a result of viral infection in asthma exacerbation in childhood. Pediatr Allergy Immunol 2002;13:47–50.

7   Simpson JL, Moric I, Wark PA, Johnston SL, Gibson PG: Use of induced sputum for the diagnosis of influenza and infections in asthma: a comparison of diagnostic techniques. J Clin Virol 2003;26:339–346.

8   Schlesinger C, Koss MN: Bronchiolitis: update 2001. Curr Opin Pulm Med 2002;8:112–116.

9   Panitch HB: Bronchiolitis in infants. Curr Opin Pediatr 2001;13:256–260.

10 Openshaw PJ, Dean GS, Culley FJ: Links between respiratory syncytial virus bronchiolitis and childhood asthma: clinical and research approaches. Pediatr Infect Dis J 2003;22:S58–S65.

11 Welliver RC: Respiratory syncytial virus and other respiratory viruses. Pediatr Infect Dis J 2003;22:S6–S12.

12 Welliver RC: Review of epidemiology and clinical risk factors for severe respiratory syncytial virus (RSV) infection. J Pediatr 2003;143:S112–S117.

13 Black CP: Systematic review of the biology and medical management of respiratory syncytial virus infection. Respir Care 2003;48:209–233.

14 Stensballe LG, Devasundaram JK, Simoes EA: Respiratory syncytial virus epidemics: the ups and downs of a seasonal virus. Pediatr Infect Dis J 2003;22:S21–S32.

15 CDC: Brief report: respiratory syncytial virus activity – United States, 2004–2005. MMWR Morb Mortal Wkly Rep 2005;54:1259–1260.

16 Thompson WW, Shay DK, Weintraub E, Brammer L, Cox N, Anderson LJ, Fukuda K: Mortality associated with influenza and respiratory syncytial virus in the United States. JAMA 2003;289:179–186.

17 CDC: Respiratory syncytial virus activity – United States, 2000–01 season. MMWR Morb Mortal Wkly Rep 2002;51:26–28.

18 CDC: Respiratory syncytial virus activity – United States, 1999–2000 season. MMWR Morb Mortal Wkly Rep 2000;49:1091–1093.

19 CDC: Update: respiratory syncytial virus activity – United States, 1998–1999 season. MMWR Morb Mortal Wkly Rep 1999;48:1104–1106, 1115.

20 CDC: Update: respiratory syncytial virus activity – United States, 1997–98 season. MMWR Morb Mortal Wkly Rep 1997;46:1163–1165.

21 Waris ME, Tsou C, Erdman DD, Zaki SR, Anderson LJ: Respiratory synctial virus infection in BALB/c mice previously immunized with formalin-inactivated virus induces enhanced pulmonary inflammatory response with a predominant Th2-like cytokine pattern. J Virol 1996;70:2852–2860.

22 Dudas RA, Karron RA: Respiratory syncytial virus vaccines. Clin Microbiol Rev 1998;11:430–439.

23 Crowcroft NS, Cutts F, Zambon MC: Respiratory syncytial virus: an underestimated cause of respiratory infection, with prospects for a vaccine. Commun Dis Public Health 1999;2:234–241.

24 Lemanske RF Jr: The childhood origins of asthma (COAST) study. Pediatr Allergy Immunol 2002;13(suppl 15):38–43.

25 Graham BS, Johnson TR, Peebles RS: Immune-mediated disease pathogenesis in respiratory syncytial virus infection. Immunopharmacology 2000;48:237–247.

26 Hussell T, Openshaw PJ: IL-12-activated NK cells reduce lung eosinophilia to the attachment protein of respiratory syncytial virus but do not enhance the severity of illness in CD8 T cell-immunodeficient conditions. J Immunol 2000;165:7109–7115.

27 Fawaz LM, Sharif-Askari E, Menezes J: Up-regulation of NK cytotoxic activity via IL-15 induction by different viruses: a comparative study. J Immunol 1999;163:4473–4480.

28 Chang J, Srikiatkhachorn A, Braciale TJ: Visualization and characterization of respiratory syncytial virus F-specific CD8(+) T cells during experimental virus infection. J Immunol 2001;167:4254–4260.

29 Mbawuike IN, Wells J, Byrd R, Cron SG, Glezen WP, Piedra PA: HLA-restricted CD8+ cytotoxic T lymphocyte, interferon-gamma, and interleukin-4 responses to respiratory syncytial virus infection in infants and children. J Infect Dis 2001;183:687–696.

30 Aung S, Tang YW, Graham BS: Interleukin-4 diminishes CD8(+) respiratory syncytial virus-specific cytotoxic T-lymphocyte activity in vivo. J Virol 1999;73:8944–8949.

31 Tripp RA, Moore D, Jones L, Sullender W, Winter J, Anderson LJ: Respiratory syncytial virus G and/or SH protein alters Th1 cytokines, natural killer cells, and neutrophils responding to pulmonary infection in BALB/c mice. J Virol 1999;73:7099–7107.

32 Chang J, Braciale TJ: Respiratory syncytial virus infection suppresses lung CD8+ T-cell effector activity and peripheral CD8+ T-cell memory in the respiratory tract. Nat Med 2002;8:54–60.

33 Chang J, Choi SY, Jin HT, Sung YC, Braciale TJ: Improved effector activity and memory CD8 T cell development by IL-2 expression during experimental respiratory syncytial virus infection. J Immunol 2004;172:503–508.

34 Boyton RJ, Openshaw PJ: Pulmonary defences to acute respiratory infection. Br Med Bull 2002;61:1–12.

35 Price JF: Acute and long-term effects of viral bronchiolitis in infancy. Lung 1990;168(suppl):414–421.

36 McBride JT: Pulmonary function changes in children after respiratory syncytial virus infection in infancy. J Pediatr 1999;135:28–32.

37 Prince AM, Jacobs RF: Prevention of respiratory syncytial virus infection in high risk infants. J Ark Med Soc 2001;98:115–118.

38 Teixeira MM, Almeida IC, Gazzinelli RT: Introduction: innate recognition of bacteria and protozoan parasites. Microbes Infect 2002;4:883–886.

39 Philpott DJ, Girardin SE: The role of Toll-like receptors and Nod proteins in bacterial infection. Mol Immunol 2004;41:1099–1108.

40 Akira S, Hemmi H: Recognition of pathogen-associated molecular patterns by TLR family. Immunol Lett 2003;85:85–95.

41 Sandor F, Buc M: Toll-like receptors. I. Structure, function and their ligands. Folia Biol (Praha) 2005;51:148–157.

42 Roeder A, Kirschning CJ, Rupec RA, Schaller M, Weindl G, Korting HC: Toll-like receptors as key mediators in innate antifungal immunity. Med Mycol 2004;42:485–498.

43 Haynes LM, Moore DD, Kurt-Jones EA, Finberg RW, Anderson LJ, Tripp RA: Involvement of toll-like receptor 4 in innate immunity to respiratory syncytial virus. J Virol 2001;75:10730–10737.

44 Haeberle HA, Takizawa R, Casola A, Brasier AR, Dieterich HJ, Van Rooijen N, Gatalica Z, Garofalo RP: Respiratory syncytial virus-induced activation of nuclear factor-kappaB in the lung involves alveolar macrophages and toll-like receptor 4-dependent pathways. J Infect Dis 2002;186:1199–1206.

45 Rallabhandi P, Bell J, Boukhvalova MS, Medvedev A, Lorenz E, Arditi M, Hemming VG, Blanco JC, Segal DM, Vogel SN: Analysis of TLR4 polymorphic variants: new insights into TLR4/MD-2/CD14 stoichiometry, structure, and signaling. J Immunol 2006;177:322–332.

46 Kadowaki N, Ho S, Antonenko S, Malefyt RW, Kastelein RA, Bazan F, Liu YJ: Subsets of human dendritic cell precursors express different toll-like receptors and respond to different microbial antigens. J Exp Med 2001;194:863–869.

47 Spisek R, Bretaudeau L, Barbieux I, Meflah K, Gregoire M: Standardized generation of fully mature p70 IL-12 secreting monocyte-derived dendritic cells for clinical use. Cancer Immunol Immunother 2001;50:417–427.

48 Rudd BD, Smit JJ, Flavell RA, Alexopoulou L, Schaller MA, Gruber A, Berlin AA, Lukacs NW: Deletion of TLR3 alters the pulmonary immune environment and mucus production during respiratory syncytial virus infection. J Immunol 2006;176:1937–1942.

49 Diebold SS, Kaisho T, Hemmi H, Akira S, Reis e Sousa C: Innate antiviral responses by means of TLR7-mediated recognition of single-stranded RNA. Science 2004;303:1529–1531.

50 Heil F, Hemmi H, Hochrein H, Ampenberger F, Kirschning C, Akira S, Lipford G, Wagner H, Bauer S: Species-specific recognition of single-stranded RNA via toll-like receptor 7 and 8. Science 2004;303:1526–1529.

51 Lund JM, Alexopoulou L, Sato A, Karow M, Adams NC, Gale NW, Iwasaki A, Flavell RA: Recognition of single-stranded RNA viruses by Toll-like receptor 7. Proc Natl Acad Sci USA 2004;101:5598–5603.

52 Yoneyama M, Kikuchi M, Natsukawa T, Shinobu N, Imaizumi T, Miyagishi M, Taira K, Akira S, Fujita T: The RNA helicase RIG-I has an essential function in double-stranded RNA-induced innate antiviral responses. Nat Immunol 2004;5:730–737.

53 Blanco-Quiros A, Gonzalez H, Arranz E, Lapena S: Decreased interleukin-12 levels in umbilical cord blood in children who developed acute bronchiolitis. Pediatr Pulmonol 1999;28:175–180.

54 Bloom ET, Horvath JA: Cellular and molecular mechanisms of the IL-12-induced increase in allospecific murine cytolytic T cell activity. Implications for the age-related decline in CTL. J Immunol 1994;152:4242–4254.

55 Chehimi J, Valiante NM, D'Andrea A, Rengaraju M, Rosado Z, Kobayashi M, Perussia B, Wolf SF, Starr SE, Trinchieri G: Enhancing effect of natural killer cell stimulatory factor (NKSF/interleukin-12) on cell-mediated cytotoxicity against tumor-derived and virus-infected cells. Eur J Immunol 1993;23:1826–1830.

56 Orange JS, Wolf SF, Biron CA: Effects of IL-12 on the response and susceptibility to experimental viral infections. J Immunol 1994;152:1253–1264.

57 Moller GM, Overbeek SE, Van Helden-Meeuwsen CG, Van Haarst JM, Prens EP, Mulder PG, Postma DS, Hoogsteden HC: Increased numbers of dendritic cells in the bronchial mucosa of atopic asthmatic patients: downregulation by inhaled corticosteroids. Clin Exp Allergy 1996;26: 517–524.

58 Jahnsen FL, Moloney ED, Hogan T, Upham JW, Burke CM, Holt PG: Rapid dendritic cell recruitment to the bronchial mucosa of patients with atopic asthma in response to local allergen challenge. Thorax 2001;56:823–826.

59 Lambrecht BN, Salomon B, Klatzmann D, Pauwels RA: Dendritic cells are required for the development of chronic eosinophilic airway inflammation in response to inhaled antigen in sensitized mice. J Immunol 1998;160:4090–4097.

60 Webb TJ, Sumpter TL, Thiele AT, Swanson KA, Wilkes DS: The phenotype and function of lung dendritic cells. Crit Rev Immunol 2005;25:465–491.

61 Julia V, Hessel EM, Malherbe L, Glaichenhaus N, O'Garra A, Coffman RL: A restricted subset of dendritic cells captures airborne antigens and remains able to activate specific T cells long after antigen exposure. Immunity 2002;16:271–283.

62 de Heer HJ, Hammad H, Soullie T, Hijdra D, Vos N, Willart MA, Hoogsteden HC, Lambrecht BN: Essential role of lung plasmacytoid dendritic cells in preventing asthmatic reactions to harmless inhaled antigen. J Exp Med 2004;200:89–98.

63 Yoneyama H, Matsuno K, Toda E, Nishiwaki T, Matsuo N, Nakano A, Narumi S, Lu B, Gerard C, Ishikawa S, Matsushima K: Plasmacytoid DCs help lymph node DCs to induce anti-HSV CTLs. J Exp Med 2005;202:425–435.

64 Megjugorac NJ, Young HA, Amrute SB, Olshalsky SL, Fitzgerald-Bocarsly P: Virally stimulated plasmacytoid dendritic cells produce chemokines and induce migration of T and NK cells. J Leukoc Biol 2004;75:504–514.

65 Kohrgruber N, Groger M, Meraner P, Kriehuber E, Petzelbauer P, Brandt S, Stingl G, Rot A, Maurer D: Plasmacytoid dendritic cell recruitment by immobilized CXCR3 ligands. J Immunol 2004;173:6592–6602.

66 Hochrein H, Schlatter B, O'Keeffe M, Wagner C, Schmitz F, Schiemann M, Bauer S, Suter M, Wagner H: Herpes simplex virus type-1 induces IFN-alpha production via Toll-like receptor 9-dependent and -independent pathways. Proc Natl Acad Sci USA 2004;101:11416–11421.

67   Tripp RA, Moore D, Anderson LJ: TH(1)- and TH(2)-TYPE cytokine expression by activated t lymphocytes from the lung and spleen during the inflammatory response to respiratory syncytial virus. Cytokine 2000;12:801–807.

68   Bartz H, Buning-Pfaue F, Turkel O, Schauer U: Respiratory syncytial virus induces prostaglandin E2, IL-10 and IL-11 generation in antigen presenting cells. Clin Exp Immunol 2002;129:438–445.

69   Bartz H, Turkel O, Hoffjan S, Rothoeft T, Gonschorek A, Schauer U: Respiratory syncytial virus decreases the capacity of myeloid dendritic cells to induce interferon-gamma in naive T cells. Immunology 2003;109:49–57.

70   Kondo Y, Matsuse H, Machida I, Kawano T, Saeki S, Tomari S, Obase Y, Fukushima C, Kohno S: Regulation of mite allergen-pulsed murine dendritic cells by respiratory syncytial virus. Am J Respir Crit Care Med 2004;169:494–498.

71   de Graaff PM, de Jong EC, van Capel TM, van Dijk ME, Roholl PJ, Boes J, Luytjes W, Kimpen JL, van Bleek GM: Respiratory syncytial virus infection of monocyte-derived dendritic cells decreases their capacity to activate CD4 T cells. J Immunol 2005;175:5904–5911.

72   Smit JJ, Rudd BD, Lukacs NW: Plasmacytoid dendritic cells inhibit pulmonary immunopathology and promote clearance of respiratory syncytial virus. J Exp Med 2006;203:1153–1159.

73   Gill MA, Palucka AK, Barton T, Ghaffar F, Jafri H, Banchereau J, Ramilo O: Mobilization of plasmacytoid and myeloid dendritic cells to mucosal sites in children with respiratory syncytial virus and other viral respiratory infections. J Infect Dis 2005;191:1105–1115.

74   Ito T, Wang YH, Liu YJ: Plasmacytoid dendritic cell precursors/type I interferon-producing cells sense viral infection by Toll-like receptor (TLR) 7 and TLR9. Springer Semin Immunopathol 2005;26:221–229.

75   Bonville CA, Rosenberg HF, Domachowske JB: Macrophage inflammatory protein-1alpha and RANTES are present in nasal secretions during ongoing upper respiratory tract infection. Pediatr Allergy Immunol 1999;10:39–44.

76   Abu-Harb M, Bell F, Finn A, Rao WH, Nixon L, Shale D, Everard ML: IL-8 and neutrophil elastase levels in the respiratory tract of infants with RSV bronchiolitis. Eur Respir J 1999;14: 139–143.

77   Gern JE, Martin MS, Anklam KA, Shen K, Roberg KA, Carlson-Dakes KT, Adler K, Gilbertson-White S, Hamilton R, Shult PA, Kirk CJ, Da Silva DF, Sund SA, Kosorok MR, Lemanske RF Jr: Relationships among specific viral pathogens, virus-induced interleukin-8, and respiratory symptoms in infancy. Pediatr Allergy Immunol 2002;13:386–393.

78   Matsuda K, Tsutsumi H, Sone S, Yoto Y, Oya K, Okamoto Y, Ogra PL, Chiba S: Characteristics of IL-6 and TNF-alpha production by respiratory syncytial virus-infected macrophages in the neonate. J Med Virol 1996;48:199–203.

79   Panuska JR, Merolla R, Rebert NA, Hoffmann SP, Tsivitse P, Cirino NM, Silverman RH, Rankin JA: Respiratory syncytial virus induces interleukin-10 by human alveolar macrophages. Suppression of early cytokine production and implications for incomplete immunity. J Clin Invest 1995;96: 2445–2453.

80   Becker S, Quay J, Soukup J: Cytokine (tumor necrosis factor, IL-6, and IL-8) production by respiratory syncytial virus-infected human alveolar macrophages. J Immunol 1991;147:4307–4312.

81   Midulla F, Huang YT, Gilbert IA, Cirino NM, McFadden ER Jr, Panuska JR: Respiratory syncytial virus infection of human cord and adult blood monocytes and alveolar macrophages. Am Rev Respir Dis 1989;140:771–777.

82   Becker S, Reed W, Henderson FW, Noah TL: RSV infection of human airway epithelial cells causes production of the beta-chemokine RANTES. Am J Physiol 1997;272:L512–L520.

83   Olszewska-Pazdrak B, Casola A, Saito T, Alam R, Crowe SE, Mei F, Ogra PL, Garofalo RP: Cell-specific expression of RANTES, MCP-1, and MIP-1alpha by lower airway epithelial cells and eosinophils infected with respiratory syncytial virus. J Virol 1998;72:4756–4764.

84   Saito T, Deskin RW, Casola A, Haeberle H, Olszewska B, Ernst PB, Alam R, Ogra PL, Garofalo R: Respiratory syncytial virus induces selective production of the chemokine RANTES by upper airway epithelial cells. J Infect Dis 1997;175:497–504.

85   Mastronarde JG, He B, Monick MM, Mukaida N, Matsushima K, Hunninghake GW: Induction of interleukin (IL)-8 gene expression by respiratory syncytial virus involves activation of nuclear factor (NF)-kappa B and NF-IL-6. J Infect Dis 1996;174:262–267.

86 Noah TL, Becker S: Chemokines in nasal secretions of normal adults experimentally infected with respiratory syncytial virus. Clin Immunol 2000;97:43–49.

87 Garofalo RP, Patti J, Hintz KA, Hill V, Ogra PL, Welliver RC: Macrophage inflammatory protein-1 alpha (not T helper type 2 cytokines) is associated with severe forms of respiratory syncytial virus bronchiolitis. J Infect Dis 2001;184:393–399.

88 Domachowske JB, Bonville CA, Gao JL, Murphy PM, Easton AJ, Rosenberg HF: MIP-1alpha is produced but it does not control pulmonary inflammation in response to respiratory syncytial virus infection in mice. Cell Immunol 2000;206:1–6.

89 John AE, Berlin AA, Lukacs NW: Respiratory syncytial virus-induced CCL5/RANTES contributes to exacerbation of allergic airway inflammation. Eur J Immunol 2003;33:1677–1685.

90 Tekkanat KK, Maassab H, Miller A, Berlin AA, Kunkel SL, Lukacs NW: RANTES (CCL5) production during primary respiratory syncytial virus infection exacerbates airway disease. Eur J Immunol 2002;32:3276–3284.

91 Miller AL, Strieter RM, Gruber AD, Ho SB, Lukacs NW: CXCR2 regulates respiratory syncytial virus-induced airway hyperreactivity and mucus overproduction. J Immunol 2003;170:3348–3356.

92 Haeberle HA, Kuziel WA, Dieterich HJ, Casola A, Gatalica Z, Garofalo RP: Inducible expression of inflammatory chemokines in respiratory syncytial virus-infected mice: role of MIP-1alpha in lung pathology. J Virol 2001;75:878–890.

93 Tripp RA, Jones L, Anderson LJ: Respiratory syncytial virus G and/or SH glycoproteins modify CC and CXC chemokine mRNA expression in the BALB/c mouse. J Virol 2000;74:6227–6229.

94 Miller AL, Bowlin TL, Lukacs NW: Respiratory syncytial virus-induced chemokine production: linking viral replication to chemokine production in vitro and in vivo. J Infect Dis 2004;189: 1419–1430.

95 Thomas MS, Kunkel SL, Lukacs NW: Differential role of IFN-gamma-inducible protein 10 kDa in a cockroach antigen-induced model of allergic airway hyperreactivity: systemic versus local effects. J Immunol 2002;169:7045–7053.

96 Bochner BS, Hudson SA, Xiao HQ, Liu MC: Release of both CCR4-active and CXCR3-active chemokines during human allergic pulmonary late-phase reactions. J Allergy Clin Immunol 2003;112:930–934.

97 Medoff BD, Sauty A, Tager AM, Maclean JA, Smith RN, Mathew A, Dufour JH, Luster AD: IFN-gamma-inducible protein 10 (CXCL10) contributes to airway hyperreactivity and airway inflammation in a mouse model of asthma. J Immunol 2002;168:5278–5286.

98 Proost P, Vynckier AK, Mahieu F, Put W, Grillet B, Struyf S, Wuyts A, Opdenakker G, Van Damme J: Microbial Toll-like receptor ligands differentially regulate CXCL10/IP-10 expression in fibroblasts and mononuclear leukocytes in synergy with IFN-gamma and provide a mechanism for enhanced synovial chemokine levels in septic arthritis. Eur J Immunol 2003;33:3146–3153.

99 Matsushima H, Yamada N, Matsue H, Shimada S: TLR3-, TLR7-, and TLR9-mediated production of proinflammatory cytokines and chemokines from murine connective tissue type skin-derived mast cells but not from bone marrow-derived mast cells. J Immunol 2004;173:531–541.

100 Rudd BD, Burstein E, Duckett CS, Li X, Lukacs NW: Differential role for TLR3 in respiratory syncytial virus-induced chemokine expression. J Virol 2005;79:3350–3357.

101 Maddox L, Schwartz DA: The pathophysiology of asthma. Annu Rev Med 2002;53:477–498.

102 Hartert TV, Peebles RS: Epidemiology of asthma: the year in review. Curr Opin Pulm Med 2000;6:4–9.

103 Arruda LK, Chapman MD: The role of cockroach allergens in asthma. Curr Opin Pulm Med 2001;7:14–19.

104 Baumholtz MA, Parish LC, Witkowski JA, Nutting WB: The medical importance of cockroaches. Int J Dermatol 1997;36:90–96.

105 Rosenstreich DL, Eggleston P, Kattan M, Baker D, Slavin RG, Gergen P, Mitchell H, McNiff-Mortimer K, Lynn H, Ownby D, Malveaux F: The role of cockroach allergy and exposure to cockroach allergen in causing morbidity among inner-city children with asthma. N Engl J Med 1997;336:1356–1363.

106 Cohn L, Ray A: T-helper type 2 cell-directed therapy for asthma. Pharmacol Ther 2000;88: 187–196.

107 Barnes PJ: Th2 cytokines and asthma: an introduction. Respir Res 2001;2:64–65.

108 Robinson DS: Th-2 cytokines in allergic disease. Br Med Bull 2000;56:956–968.
109 Lee SY, Kim SJ, Kwon SS, Kim YK, Kim KH, Moon HS, Song JS, Park SH: Distribution and cytokine production of CD4 and CD8 T-lymphocyte subsets in patients with acute asthma attacks. Ann Allergy Asthma Immunol 2001;86:659–664.
110 O'Sullivan S, Cormican L, Faul JL, Ichinohe S, Johnston SL, Burke CM, Poulter LW: Activated, cytotoxic CD8(+) T lymphocytes contribute to the pathology of asthma death. Am J Respir Crit Care Med 2001;164:560–564.
111 Schwarze J, Makela M, Cieslewicz G, Dakhama A, Lahn M, Ikemura T, Joetham A, Gelfand EW: Transfer of the enhancing effect of respiratory syncytial virus infection on subsequent allergic airway sensitization by T lymphocytes. J Immunol 1999;163:5729–5734.
112 Braciale TJ: Respiratory syncytial virus and T cells: interplay between the virus and the host adaptive immune system. Proc Am Thorac Soc 2005;2:141–146.
113 Schaller MA, Lundy SK, Huffnagle GB, Lukacs NW: CD8(+) T cell contributions to allergen induced pulmonary inflammation and airway hyperreactivity. Eur J Immunol 2005;35:2061–2070.
114 Dakhama A, Park JW, Taube C, Joetham A, Balhorn A, Miyahara N, Takeda K, Gelfand EW: The enhancement or prevention of airway hyperresponsiveness during reinfection with respiratory syncytial virus is critically dependent on the age at first infection and IL-13 production. J Immunol 2005;175:1876–1883.
115 Stanciu LA, Shute J, Promwong C, Holgate ST, Djukanovic R: Increased levels of IL-4 in CD8+ T cells in atopic asthma. J Allergy Clin Immunol 1997;100:373–378.
116 Erb KJ, Le Gros G: The role of Th2 type CD4+ T cells and Th2 type CD8+ T cells in asthma. Immunol Cell Biol 1996;74:206–208.
117 Folkard SG, Westwick J, Millar AB: Production of interleukin-8, RANTES and MCP-1 in intrinsic and extrinsic asthmatics. Eur Respir J 1997;10:2097–2104.
118 Alam R, York J, Boyars M, Stafford S, Grant JA, Lee J, Forsythe P, Sim T, Ida N: Increased MCP-1, RANTES, and MIP-1alpha in bronchoalveolar lavage fluid of allergic asthmatic patients. Am J Respir Crit Care Med 1996;153:1398–1404.
119 Holgate ST, Bodey KS, Janezic A, Frew AJ, Kaplan AP, Teran LM: Release of RANTES, MIP-1 alpha, and MCP-1 into asthmatic airways following endobronchial allergen challenge. Am J Respir Crit Care Med 1997;156:1377–1383.
120 Blease K, Mehrad B, Standiford TJ, Lukacs NW, Kunkel SL, Chensue SW, Lu B, Gerard CJ, Hogaboam CM: Airway remodeling is absent in CCR1–/– mice during chronic fungal allergic airway disease. J Immunol 2000;165:1564–1572.
121 Ma B, Zhu Z, Homer RJ, Gerard C, Strieter R, Elias JA: The C10/CCL6 chemokine and CCR1 play critical roles in the pathogenesis of IL-13-induced inflammation and remodeling. J Immunol 2004;172:1872–1881.

Nicholas W. Lukacs, PhD
109 Zina Pitcher
4059 BSRB
Ann Arbor, MI 48109-2200 (USA)
Tel. +1 734 764 5135, E-Mail nlukacs@umich.edu

Sjöbring U, Taylor JD (eds): Models of Exacerbations in Asthma and COPD.
Contrib Microbiol. Basel, Karger, 2007, vol 14, pp 83–100

Lipopolysaccharide Challenge of Humans as a Model for Chronic Obstructive Lung Disease Exacerbations

Sergei A. Kharitonov[a,b], *Ulf Sjöbring*[c,d]

[a]Section of Airway Disease, National Heart and Lung Institute, Imperial College London and [b]Royal Brompton and Harefield NHS Trust, London, UK; [c]Department of Medial Science, AstraZeneca R&D and [d]Institute of Laboratory Medicine, Lund University, Lund, Sweden

Abstract

Endotoxin, or lipopolysaccharide (LPS), is a constituent of the outer cell membrane of Gram-negative bacteria. LPS is a highly potent proinflammatory substance, that, when inhaled, dose-dependently causes fever, chills, and bronchoconstriction. These symptoms are accompanied by a proinflammatory response in sputum and bronchoalveolar lavage fluid with elevation of neutrophils, macrophages and certain cytokines/chemokines. This response can be partially modified with certain drugs. Similar inflammatory changes are observed both in the stable state of chronic obstructive lung disease (COPD) and during exacerbations of this disease. Cigarette smoke, which contains bioactive LPS, is the most common cause of COPD and may also precipitate exacerbations. In addition, the presence of Gram-negative bacteria in the lower airways is a distinguishing feature both of stable COPD and of exacerbations. Based on this knowledge we argue here that inhaled LPS provocation of healthy volunteers can be used as a model or COPD as well as for exacerbations of this disease.

Introduction

Endotoxin, also designated lipopolysaccharide (LPS), is a heterogeneous constituent of the outer cell membrane of Gram-negative bacteria [1, 2]. LPS is continuously shed into the environment and acts as a highly potent proinflammatory substance. In humans, acute inhalation of LPS dose-dependently causes clinical symptoms, including fever, chills, and bronchoconstriction. These symptoms are accompanied by a proinflammatory response with increases of neutrophils, macrophages and certain chemokines/cytokines in sputum and bronchoalveolar lavage fluid (BALF) [3–6]. The increases are accompanied by elevation of the corresponding cells in blood, as well as of acute-phase plasma proteins such as IL-6, C-reactive protein (CRP), LPS-binding protein (LBP) and E-selectin.

Chronic exposure of LPS in organic dust is believed to contribute to the development of chronic bronchitis and impairment of lung function that can be observed in certain occupational settings (cotton workers, wool processing workers, workers in recycling plants) [7, 8]. Moreover, it has been suggested that LPS in cigarette smoke can be a driver of disease in chronic obstructive pulmonary disease (COPD). In addition, since the lower respiratory tract of patients with COPD is often chronically colonized with Gram-negative bacteria [9], it may be considered if this flora contributes to disease progression in COPD [10–12]. Finally, since the same Gram-negative species that colonize the airways are also implicated in exacerbations of COPD, one may hypothesize that LPS released from rapidly dividing bacteria may contribute to the development of exacerbations. Against this background it is not surprising that medications targeting COPD have been tested in healthy volunteers or patients after a challenge with inhaled LPS, using an elevation of inflammatory cells and cyto-/chemokines in the airways as endpoints.

Here we summarize the results obtained in acute LPS challenge studies in man, with an emphasis on provocation via the inhaled route.

LPS and Lipooligosaccharide

LPS is an extremely potent stimulator of the immune system, active at femtomole levels [1, 2, 13]. It is a large molecule consisting of a lipid and a polysaccharide joined by a covalent bond. LPS forms a major component of the outer membrane of Gram-negative bacteria contributing to their structural integrity and protecting the membrane from certain kinds of chemical attack. It may also function as an adhesion molecule. The molecule comprises three distinct parts: polysaccharide side chains, a core oligosaccharide and lipid A. Lipid A contains unusual fatty acids and is embedded in the outer membrane, while the rest of the LPS projects from the bacterial surface. The core oligosaccharide is attached to lipid A and contains unusual sugars such as keto-deoxyoctulonate. The polysaccharide side chain (or the O-antigen) is easily recognized by antibodies and not surprisingly the O side chain is highly variable. Consequently, the composition of the O side chain not only varies between different Gram-negative bacterial strains but also between serotypes within a single species.

Most of the toxicity of Gram-negative bacteria can be associated with the lipid A part of the molecule, since it is only this region that binds to/evokes signals via the Toll-like receptor 4 (TLR4; see below). Since the lipid A moiety is not exposed on intact bacteria, LPS has to be released to the environment to create damage. Such release generally occurs when bacteria are dividing rapidly or when they are attacked by the host immune system. It should be noted that although lipid A is the most conserved part of LPS it does show some variability which is also reflected in the variable ability of different LPS variants to evoke a host response through TLR4. There is evidence that different lipid A structures are associated with human disease. Thus, *Pseudomonas aeruginosa* strains isolated from cystic fibrosis patients have lipid A structures that are not found in environmental isolates [14]. On the other hand evidence from the *Yersinia pestis* system suggests that highly virulent Gram-negative pathogens express lipid A variants that are poorly recognized by the human TLR4, thereby allowing evasion of the immune response [15]. The variable host response to different LPS/lipooligosaccharide (LOS) variants needs to be considered both when choosing LPS for challenge and when discussing the role of this molecule in the pathogenesis of COPD. Thus, *Haemophilus influenzae* and *Moraxella catharralis*, species that are of particular relevance in the context of COPD, express LOS instead of LPS [16]. LOS is analogous to LPS and shares a similar lipid A structure with an identical array of functional activities as LPS, but lacks O-antigen units and its oligosaccharide is limited to 10 saccharide units.

LPS-Binding and Signalling Mechanisms

TLR4 appears to be the major, and possibly the only, cellular receptor for LPS. Null mutations of TLR4 in mice ablate the response to LPS, and humans with certain single-nucleotide polymorphisms in the TLR-4 gene may demonstrate hyporesponsivess to inhaled LPS [5, 17]. TLR4 is highly expressed on white blood cells (monocytes, neutrophils, dendritic cells), but can also be found on endothelial cells and on the lung epithelium.

LPS first binds to LBP, a circulating opsonin for membrane-anchored CD14. LPS, LBP and CD14 form a ternary complex at the membrane. A secreted protein, MD2, binds to the extracellular domain of TLR4, where it facilitates LPS responsiveness, possibly by stabilizing TLR4 dimers. The stabilized receptor complex signals via Toll homology domains through the adaptor protein myeloid differentiation factor MyD88. The N-terminal death domain of MyD88 undergoes homophilic interaction with the death domain of the serine kinase IL-1 receptor-associated kinase-1 (IRAK-1) that is then autophosphorylated and forms a complex with TRAF6, ultimately leading to IκB activation and to NF-κB translocation to the nucleus. Recently, a second, MyD88-independent, signalling route for TLR4 was discovered involving the adaptor proteins TRIF (Toll/IL-1R domain-containing adaptor-inducing IFN-β) and TRAM (TRIF adaptor-related adaptor molecule). This pathway initiates the type 1 IFN response as well as a late activation of NF-κB. Together, these signalling mechanisms will result in the release of a number of proinflammatory cytokines such as TNF-α, IL-6 and α-, β- and λ-interferons as well as of chemokines including IL-8, MIP-1α and MCP-1 [17–20].

In this context it is worth mentioning that exposure of macrophages to LPS induces a hyporesponsive state to a second challenge with the molecule. This tolerance may serve to limit, but not completely abrogate, the response to chronic LPS exposure [21, 22]. Apart from tolerance LPS bioactivity can be limited by several factors including bactericidal/permeability-increasing protein that binds LPS or by cellular uptake [23]. The effects of smoking on LPS inactivation and clearance are unknown.

Inhaled LPS – Links to Disease

Consequences of Occupational LPS Exposure

Workers chronically exposed to cotton or grain dust have an increased prevalence of cough and phlegm, as well as an annual decrease in lung function compared to that of controls. These signs and symptoms have been associated with LPS on dust particles; in particular the concentration of LPS in the inhaled dust is more critical than the level of dust particles [7].

The epidemiological evidence is supported by experimental findings. In vitro LPS-containing dust particles induce alveolar macrophages to release factors chemotactic for neutrophils. In in vivo models, using LPS-responding (TLR4-positive) and non-responding (TLR4-deficient) mice, subchronic inhalation of grain dust extract resulted in the development of chronic airway disease only in TLR4-positive mice. At relevant exposure levels, repeated challenge of animals, as well as of healthy individuals, with LPS or corn dust extract resulted in similar symptoms and changes in airflow accompanied by increases in BAL inflammatory cells and mediators [7].

LPS in Tobacco Smoke

There is evidence that cigarette tobacco contains high concentrations of LPS and that the particulate phase of cigarette smoke contains sufficient levels of biologically active LPS to cause chronic bronchitis. Using the Limulus amoebocyte lysate assay, bioactive LPS was detected in the tobacco and filter tip components of unsmoked cigarettes, as well as on smoke particles generated with an automated smoking machine. Bioactive LPS was detected in both main- and sidestream smoke [24]. It has been shown that each smoked cigarette can deliver 17.4 pmol of endotoxin [25], and it can be estimated that smoking one pack of cigarettes per day delivers a dose of LPS that is comparable to the levels of LPS associated with adverse health effects in cotton textile workers [24]. Cigarette smoke is also a major cause of increased levels of LPS in indoor environments [26]. The medical consequences of this exposure are unknown.

However, it is fair to say that although systemic absorption of inhaled LPS may occur, no differences in the blood LPS levels were detected in smokers compared to non-smokers [24]. The reason for this is unclear, but may be due to a fast binding of LPS to circulating proteins (such as LBP and bactericidal/ permeability-increasing protein) or to cells.

In summary, the fact that cigarette smoke contains large amounts of LPS may contribute to the high prevalence of respiratory disorders among smokers and may also be involved in the pulmonary health hazards caused by environmental smoke.

Chronic Bacterial Colonization of the Lower Respiratory Tract

The lower respiratory tract (below the vocal cords) is sterile in healthy individuals. However, a sizeable fraction of COPD patients are colonized in their lower respiratory tract with Gram-negative pathogens with relatively low virulence, such as *M. catarrhalis* and non-encapsulated (non-typable) *H. influenzae* [9]. Some individuals are also carriers of *Streptococcus pneumoniae*, which in general terms can be regarded as somewhat more virulent. Surprisingly little is known about the subspecies, serotypes and the degree of encapsulation of the *S. pneumoniae* strains isolated from patients with COPD. *H. influenzae* is by far the most common colonizer, and can be cultured from 15–40% of the patients [27, 28].

Analyses of sputum and BAL have suggested that otherwise healthy cigarette smokers can also be colonized. However, more stringent analyses using protected brush specimens have not been able to confirm this [29]. In fact, available data suggest that colonization becomes increasingly common as the disease progresses. It could therefore be argued that bacterial colonization may be a bystander phenomenon. However, this appears less likely since it has been described that colonization is (1) linked to a faster decline in COPD and (2) associated with elevated markers of systemic inflammation [30, 31]. From a mechanistic point of view, and as already pointed out, it should be emphasized that *H. influenzae* and *M. catharralis* both produce a LOS with a lipid A moiety. It is possible that colonizing bacteria, perhaps embedded in biofilms, have a lower rate of replication. However, even if this is the case, for the bacteria to stay on they must proliferate to withstand the continuous clearance by 'unspecific' (mucus clearance), innate (complement, phagocytosis) and adaptive (IgA, IgG) host defence mechanisms. Unavoidably, such proliferation is coupled to a release of LPS/LOS into the airways. Surprisingly, to our knowledge, there has been no attempt to quantitate the amount of LPS in the airways of colonized versus non-colonized individuals.

Oral Cavity

A more speculative source for LPS may be the oral cavity. Thus, it has been suggested that chronic lung disease, including COPD, may be linked to poor dental status and in particular to periodontitis, a condition in which chronic infection with Gram-negative species, such as *Porphyromonas gingivalis*, is strongly implicated [32]. Further epidemiological investigations are required to verify such a link.

Role of Bacteria in Exacerbations

Bacteria are believed to cause approximately 50% of the exacerbations, alone or following virus infection [9]. The evidence supporting this statement includes the observation that bacteria are much more frequently isolated from patients with exacerbations than from individuals with stable disease, even using rigorous methodology and that treatment with antibiotics is beneficial [33, 34]. The most commonly isolated species is again *H. influenzae* followed by *S. pneumoniae* and *M. catharralis*. Other Gram-negative species, such as Enterobacteriacae and *P. aeruginosa*, are occasionally implicated, particularly in advanced COPD.

Here, it should be emphasized that the bacteriological investigations performed so far have used conventional ($\sim$100-year-old) techniques to detect and quantitate bacteria. Only recently has a role for 'atypical' species, which do not grow under these conditions, been established in exacerbations of COPD. Thus, in one study it was estimated that about 4% of the exacerbations was caused by *Chlamydia pneumoniae* (that also produces LPS) [35]. The role of anaerobic species, many of which are LPS-producing Gram-negative ones, is completely unknown. It is therefore possible that the use of more up-to-date molecular methodologies may unravel novel LPS-producing species as drivers of pathogenesis in both stable and exacerbating COPD.

Methodology of LPS Inhalation Challenge in Humans

Technique

The LPS solution is usually prepared by dissolving *Escherichia coli* endotoxin in sterile (and endotoxin-free) 0.9% saline. Most commonly LPS prepared from the *E. coli* strain 026:B6 (Sigma Chemicals, Poole, UK) has been used.

Occasionally other strains, such as the *E. coli* 0:113 strain, have also been used as the LPS source. The same saline used to dissolve the LPS is used for control provocation [36]. Most often a breath-activated nebulizer (e.g. Mefar dosimeter MB3, Brescia, Italy) has been used for delivery. This system delivers particles with an aerodynamic mass median diameter of 3.4–4 μm. It has been shown that it is important to test the actual delivery dose per breath of a particular nebulizer before the experiment, and that the best practice is to inhale the total dose in no less than 5 breaths to minimize the effect of inadvertent mistakes by the subjects [36]. Some investigators have used an aerosol delivery system (e.g., Mallinckrodt Diagnostica, Petten, The Netherlands) whereby LPS is delivered to a large-volume collapsible reservoir of 30 litres, made of static field dissipative material (RCAS 1206, Richmond Redlands, Calif., USA) [37]. Subsequently, the entire volume of the reservoir is inhaled by tidal breathing for ~3 min through a three-way valve system with the nose clipped.

Dosing

Although a recent study in healthy volunteers showed an increase of neutrophils in induced sputum with an LPS dose as low as 5 μg [4], the most common is between 40 and 60 μg [4, 38–40]. Another approach to calculate the LPS dose is to equate to the exposure levels reached in an appropriate occupational setting [8].

There is evidence of desensitization to LPS stimulation both from in vitro and in vivo models. The fact that individuals regularly exposed to LPS-containing dust particles do not display daily acute symptoms supports the fact that desensitization also occurs in man. However, several reports have demonstrated that a slight escalation of inhaled LPS doses in the 0.5- to 50-μg interval does not appear to ablate the response [4, 41].

In this context it needs to be mentioned that dosing with shorter intervals than 1 week may be hazardous: dosing of rodents with two consecutive injections of LPS 24 h apart elicits the so-called Shwartzman reaction, characterized by systemic manifestations reminiscent of disseminated intravascular coagulation [42].

Timing of the Sampling

The timing of sampling is dependent both on the compartment that is sampled and the nature of the sample. To capture certain immunological markers in sputum, such as IL-8, sampling may optimally be done already 4–6 h after the LPS dose. In contrast, an increase in lymphocytes and some mediators will occur at a later time point (24–48 h). Similarly, while an increase in circulating neutrophils, or in IL-6, is evident 6–8 h after challenge, the change in acute phase reactants, such as CRP, will peak around 24 h [4, 39, 43]. Most sputum biomarkers have been studied at 6 or 24 h after LPS challenge.

Methods to Sample and Measure Local Biomarkers
following LPS Challenge

Up to now most studies of biomarkers have involved analysis of sputum, BALF or plasma. However, in this chapter we also list some other techniques that may be considered following LPS challenge and that are of particular interest since they are non-invasive [44].

Induced Sputum

The inflammatory process, and potential effects of LPS, can be studied relatively non-invasively using spontaneous or induced sputum [45]. Sputum induction is usually performed with repeated administration of an aerosol of hypertonic saline at concentrations between 3 and 5% [45]. After each inhalation, the subjects are asked to blow their nose, rinse their mouth with water, swallow (to minimize contamination with postnasal drip or saliva) and then cough into a clear container. The primary concern with sputum induction, particularly in patients with COPD, is safety, since inhalation of saline causes bronchoconstriction. The occurrence of bronchoconstriction can be reduced by using a relatively low-output ultrasonic nebulizer, which does not reduce the success of induction, and by the inhalation of a β_2-agonist to bronchodilate. The FEV_1 should be measured before and after each inhalation, and the inhalations should be discontinued if there is a fall of >20%. Any bronchoconstriction can be reversed by further inhalation of the bronchodilator.

It should be emphasized that sputum induction in itself can cause a transient (lasting up to 24 h) sputum neutrophilia and a reduction in sputum macrophages at 8 h [36]. Therefore, it is important not to repeat sputum induction for at least 24 h.

Exhaled Nitric Oxide

In asthma, increased exhaled nitric oxide (eNO) levels mainly derive from the larger airways. In contrast, using a novel method for measuring eNO at several exhalation flow rates it was demonstrated that excessive eNO was produced in the small airways of COPD patients, and that the increase was correlated to disease severity [46]. This is consistent with the concept that changes in the small airways are a determinant of disease progression in COPD. It has been shown that the eNO levels were elevated in healthy subjects after LPS inhalation at 1 h and the levels remained elevated until the 6th hour [47]. It remains to be established whether LPS causes signs of inflammation in the small airway in patients with COPD. If that were the case the technique would be suitable for analyzing the therapeutic potential of pharmaceutical agents.

Exhaled Temperature and Bronchial Blood Flow

Asthma is characterized by vascular hyperperfusion, which is reflected in elevated exhaled temperature, possibly as an index of airway inflammation [48]. In contrast, in COPD the exhaled temperature is low [49]. This may reflect a reduced endothelial NO release and/or an impaired pulmonary circulation. The lack of response of the bronchial blood flow after inhalation of albuterol, which has vasorelaxant properties, may be an indicator of the degree of tissue remodelling in COPD. It is therefore possible that the exhaled temperature may be used to monitor the pulmonary circulation and the extent of the ventilation/perfusion defect in patients with COPD. The effects on exhaled temperature by inhaled LPS in healthy volunteers and COPD patients remain to be established.

Exhaled Breath Condensate

Exhaled breath condensate is collected by cooling or freezing exhaled air. Although the collection procedure has not been standardized, there is clear evidence suggesting that abnormalities in the condensate composition may reflect biochemical changes in the airway lining fluid [50]. For example, higher concentrations of total protein were found in exhaled condensate from young smokers when compared to non-smokers [51]. Various proteins derived from the airways (and unlikely to be contaminated with saliva) could be detected in exhaled breath condensate by two-dimensional electrophoresis [52]. Again the impact of inhaled LPS on these parameters will need to be established.

Inhalation of LPS – Findings in Healthy Volunteers

Effects on Systemic Signs and Symptoms

There is a considerable individual variation both in the systemic and the local response to inhaled LPS. Factors that may influence the response include sex and atopy/asthma. The molecular basis for the variation is not known. However, the variation notwithstanding, both the systemic and the local response to inhaled LPS show a dose-dependent pattern. At doses at or below 5 μg the signs and symptoms of systemic reactions are few. However, when increasing the dose above 5 μg symptoms and signs of toxicity appear. The most common symptoms include chills/shivers, malaise, myalgia and fatigue. Headache, nausea and diarrhoea can also occur. These symptoms usually appear 8–12 h after dosing [4]. Signs include fever, which may be seen as early as 60 min postchallenge. The temperature usually peaks at ~7 h, with a maximum elevation up to 0.7°C, and returns to normal values after 24 h [36]. An impact on cardiovascular parameters, such as pulse rate and blood pressure, is also not uncommon [36]. It is therefore important to keep the subjects under observations with repeated hourly assessment of symptoms and vital signs, including body temperature, blood pressure and FEV_1, for at least 8 h. Subjects may then be allowed home and re-attend 24 h after the LPS challenge for re-evaluation.

Effects on Systemic Biomarkers

Inhalation of LPS (40 μg) caused a doubling of blood neutrophils and myeloperoxidase at 24 h [39]. Similarly, 50 μg of LPS caused blood neutrophilia both at 6 and 24 h although the elevation was more marked at 6 h. In contrast, for CRP, LBP and soluble E-selectin, the increases were more evident after 24 h [40, 41]. Michel et al. [4], apart from noting increased neutrophil and CRP levels (already after 5 μg), found that 50 μg of inhaled LPS caused an increase of circulating monocytes and lymphocytes. Following provocation with 50 μg increased levels of CRP could also be measured in urine [4].

Following segmental instillation of LPS, neutrophils and lymphocytes, but not monocytes, were increased in the circulation. In addition, mediators such as G-CSF, IL-6 and IL-1 receptor agonist were increased, as were the acute phase reactants CRP and serum amyloid A [43].

Whether the systemic response is due to amplification of proinflammatory signals through airway cells or whether it results from a mere leakage of LPS into plasma is not clear. The observation that inhaled LPS challenge in healthy subjects caused increased levels of the 16-kDa Clara cell protein (CC16) may support the second alternative, since the presence of this protein in the circulation may be regarded as a non-invasive test of the alveolocapillary barrier permeability [6].

Respiratory Symptoms and Effects on Lung Function

LPS inhalation causes breathlessness and dry cough in a substantial number of individuals following dosing of ≥40 μg. Infrequently, cough with phlegm and chest tightness can also occur [39].

Objectively inhalation of LPS affects lung function in both normal subjects and asthmatics. The data on dose dependency from different investigators are not entirely in agreement. Thus, Kline et al. [3] reported that a subset of healthy individuals reacted with a 20% decline in FEV_1 following 6.5 μg of inhaled LPS, whereas another subset of individuals had FEV_1 values above 90% even after 40 μg. In contrast, other investigators found that doses of LPS up to 100 μg had very limited effect on FEV_1 in normal subjects [38]. Some of these inconsistencies may be attributable to differences in the protocols, but, as mentioned previously, there is a possibility that the differences may be due to a variability in the response to LPS, including differences in TLR4 binding and/or signalling.

Effects on Local Biomarkers

Sputum
Similar to what is the case in COPD, the response to LPS provocation is dominated by an increase in neutrophils and to some extent in monocytes. Given the dominating neutrophilic response it is not surprising that there is an elevation of myeloperoxidase and matrix metalloproteinase-9, both major components of neutrophilic granules. Again, these observations are consistent with the findings in COPD. In addition, several mediators, including TNF-α, IL-6 and IL-8, have been described to be increased both following LPS challenge and in COPD [4, 39–41, 53]. There are also reports of increases of lymphocytes in sputum [39]; the different patterns observed may reflect sampling at different time points after provocation.

Bronchoalveolar Lavage
Similar to sputum, inhaled LPS caused an increase in neutrophils in BALF 3 h after inhalation, but unlike sputum there was no increase in monocytes/ macrophages or lymphocytes [54]. This is consistent with the results by Maris et al. [38], who in addition to elevation of neutrophils found that the levels of neutrophil degranulation products such as myeloperoxidase and elastase were increased. Similar to in sputum (and in COPD), mediators, such as the cytokines IL-6 and TNF-α and the chemokine IL-8, were increased [38].

LPS also appears to induce a local imbalance in coagulation versus fibrinolysis in the BALF of healthy subjects. Thus, procoagulant factors, such as the thrombin-antithrombin complex, factor VIIa, and tissue factor were increased concomitantly with an increase of the antifibrinolytic factor PAI-1 [38, 55]. Again, similar changes have been observed in sputum from patients with COPD [56].

O'Grady et al. [43] instilled healthy volunteers with LPS at 1–4 ng/kg into one lung segment and saline into the contralateral segment. This procedure was followed by a collection of BALF at 2, 6, 24 or 48 h. Endotoxin instillation resulted in a focal and systemic inflammatory response with a distinct time course. There was an early elevation (at 2–6 h) of neutrophils, and of some cytokines including TNF-α, TNF receptors, IL-1β, and IL-6 as well as of the chemokines IL-8, MCP-1, MIP-1α, and MIP-1β. In a later phase (24–48 h) neutrophils, macrophages, monocytes, and lymphocytes were all increased, whereas most of the mediators, apart from the TNF receptors and L-selectin, had returned to the basal levels. The methodology has also been used to study the effect of an anti-MCP-1 antibody on cellular influx in patients with COPD (see below) [57].

Findings in Smokers and COPD

There are only a limited number of studies involving LPS challenge in habitual smokers and COPD patients. LPS caused a rise in differential and absolute neutrophil counts and IL-8 in sputum at 6 h after challenge of both smokers and non-smokers [58]. In smokers these biomarkers remained elevated at 24 h. In addition, at 6 h after LPS inhalation, the levels of 8-isoprostane in exhaled breath condensate were higher in smokers compared to non-smokers. NF-κB activation was increased at 6 h after LPS in sputum macrophages from both healthy individuals and from smokers [58]. It is worth noting that disease severity in COPD was associated with an increased epithelial expression of NF-κB [59]. One may therefore speculate that inhaled LPS challenge will mimic an acute COPD exacerbation of bacterial origin and may induce a cascade of events resulting in NF-κB induction and activation, cytokine and chemokine production and further inflammatory cell infiltration. The recent demonstration that therapeutic intervention can modulate a relevant cellular influx induced by LPS further supports this notion [57].

Testing Drug Properties in the LPS Challenge Model

In the following we provide some examples from the literature where the LPS model has been used to study the effects of anti-COPD drugs, in healthy volunteers, smokers and COPD patients.

Salmeterol

Maris et al. [38] studied the impact of pretreatment with the inhaled β_2-agonist salmeterol (100 μg given 30 min before exposure) on the recruitment of cells and mediators in BALF following inhalation of LPS (100 μg). The measurements were performed 6 h post inhalation. Salmeterol reduced LPS-induced neutrophil influx as well as degranulation of the neutrophils, as monitored by myeloperoxidase and the drug also reduced the TNF-α levels.

These findings are at odds with a report by Wallin et al. [54] who failed to find an effect of salmeterol 50 μg twice daily for 3 weeks (last dose 1–2 h before the provocation) on the increased levels of neutrophils in BALF seen after inhaled LPS (50 μg). The reason for the discrepancy is unclear.

In another publication, Maris et al. [60] reported that salmeterol, either with or without LPS inhalation, enhanced markers of fibrinolysis in BALF, but did not influence LPS-induced changes in coagulation.

PDE4 Inhibitors

Recently, it was demonstrated that the oral PDE4 inhibitor cilomilast (15 mg, once daily, for 6 days) failed to reduce airway inflammation (in sputum) induced by inhaled LPS in healthy subjects [40]. Local measurements included TNF-α, which is known to be efficiently inhibited by cilomilast in several pre-clinical in vitro and in vivo models. LPS-induced elevation of CRP in plasma was attenuated by cilomilast but the effect was not statistically significant versus controls [40]. While the lack of effect on the local response of cilomilast was surprising it is not entirely without precedence; although PDE4 inhibitors were effective on the increase of systemic TNF-α following intravenous (i.v.) administration of LPS in rats the drug was found to be ineffective when LPS was administered intratracheally in this species [61].

Corticosteroids

Oral prednisolone (10 mg, once daily, for 6 days) significantly reduced acute systemic inflammation as measured by CRP in healthy subjects subjected to 50 μg of inhaled LPS [40]. However, this treatment also failed to reduce air-way inflammation in sputum induced by inhaled LPS [40]. In contrast, a 6-day treatment with 20 mg/day methylprednisolone was able to restore the alveolo-capillary barrier permeability defect caused by inhaled LPS, as measured by intravascular leakage of CC16 [6]. These data suggest that oral glucocorticoids reduce the systemic LPS exposure through alveolocapillary leakage, but pro-vide little benefit to the local proinflammatory reaction of inhaled endotoxin.

Anti-MCP-1

The segmental LPS instillation approach described by O'Grady et al. [43] was recently used to evaluate the effect of a systemically administered anti-MCP-1 monoclonal antibody (ABN912) on cellular recruitment to the bronchi in a group of individuals with mild COPD (Gold stage 1) [57]. The patients were treated with the antibody, or placebo, and after 10 days a segmental challenge with LPS was performed using saline challenge in another segment as control. Analysis of the cellular recruitment to BALF from the challenged segments after 24 h demonstrated that the antibody was able to attenuate the influx of mono-cytes into the airways of COPD patients, and in particular following LPS chal-lenge. These data are very encouraging, and certainly provide proof of mechanism, but will require benchmarking against clinical outcomes for validation. Once available such validation may also apply to the LPS model as such.

Other Routes of Administration

Intranasal Challenge

The focus on the lung in COPD ignores the anatomic continuity between the lower and upper airway. The epithelium reacts in a similar way to noxious stimuli and there is a correlation between the degree of inflammation in the upper and lower airways of COPD patients [62]. Indeed, COPD patients had a high prevalence of nasal symptoms (75%), and more than half reported nasal discharge (52.5%) as well as sneezing (45.9%) [63]. Although no significant relationship was found between nasal symptoms and FEV_1 in COPD, significant associations were demonstrated between nasal score and daily sputum production. There was also a trend to increased nasal symptoms in frequent exacerbators compared to infrequent exacerbators [63]. A possible mechanism for cross-talk between the upper and lower airways in COPD patients has recently been suggested, as an increased nasal concentration of IL-8 was related to the levels of this neutrophil chemoattractant in the lower airway [62]. Recently, a small increase in several inflammatory markers, including IL-8 and soluble TNF-α factor receptor 75 (sTNF-R75), in nasal lavage fluid was registered as part of the response to airborne endotoxin in cotton workers [64]. These changes were accompanied by a reduction in the lung function during a 6-week exposure. The findings may have implications for the use of the nose as a model of the lower airway in COPD patients, suggesting that local intranasally introduced LPS challenge may be a safer alternative to inhaled LPS challenge.

i.v. Administration

Administration of LPS through the i.v. route replicates many of the features of Gram-negative sepsis, including fever, release of inflammatory mediators (IL-6, IL-8, TNF-α, and many others), leucocytosis dominated by neutrophilia and an increase in acute phase proteins, such as CRP and serum amyloid A [65]. In addition i.v. LPS also induces a change in haemostasis favouring a procoagulative, antifibrinolytic state, resembling the early intravascular changes occurring in severe sepsis [66]. Endotoxin impacts the systemic circulation, inducing a hyperdynamic cardiovascular state with a reversible depression of the left ventricular function. Moreover, systemically administered LPS causes mild and reversible changes in lung function, an increased respiratory frequency, decreased inspiratory time, a widened alveolar-arterial oxygen tension gradient, dyspnoea and a decreased FEV_1 [67]. The i.v. LPS model has therefore been widely used in early clinical studies with new anti-inflammatory or antisepsis drugs. For example, eritoran (E5564), a lipid A analogue that functions as a TLR4 antagonist, reduced a number of LPS-induced systemic phenomena, including elevation of temperature, chills, myalgia, tachycardia, and increased levels of CRP, white blood cells and plasma cytokines [68].

Conclusions

Cigarette smoke, which contains LPS, is the most common cause of COPD and may also precipitate exacerbation of this disease. In addition, Gram-negative bacterial presence in the lower airways is a distinguishing feature of both stable and exacerbating COPD. LPS induces similar inflammatory changes in sputum and BALF as can be observed both in stable and exacerbating COPD, including elevation of neutrophils and mediators typically controlled through the TLR4 signalling pathway. Findings in early clinical trials support the utility of the LPS model, demonstrating that certain compounds can modify the response to inhaled LPS.

References

1   Bishop RE: Fundamentals of endotoxin structure and function. Contrib Microbiol. Basel, Karger, 2005, vol 12, pp 1–27.
2   Trent MS, Stead CM, Tran AX, Hankins JV: Diversity of endotoxin and its impact on pathogenesis. J Endotoxin Res 2006;12:205–223.
3   Kline JN, Cowden JD, Hunninghake GW, Schutte BC, Watt JL, Wohlford-Lenane CL, Powers LS, Jones MP, Schwartz DA: Variable airway responsiveness to inhaled lipopolysaccharide. Am J Respir Crit Care Med 1999;160:297–303.
4   Michel O, Nagy AM, Schroeven M, Duchateau J, Neve J, Fondu P, Sergysels R: Dose-response relationship to inhaled endotoxin in normal subjects. Am J Respir Crit Care Med 1997;156:1157–1164.
5   Michel O, LeVan TD, Stern D, Dentener M, Thorn J, Gnat D, Beijer ML, Cochaux P, Holt PG, Martinez FD, Rylander R: Systemic responsiveness to lipopolysaccharide and polymorphisms in the toll-like receptor 4 gene in human beings. J Allergy Clin Immunol 2003;112:923–929.
6   Michel O, Murdoch R, Bernard A: Inhaled LPS induces blood release of Clara cell specific protein (CC16) in human beings. J Allergy Clin Immunol 2005;115:1143–1147.
7   Balmes J, Becklake M, Blanc P, Henneberger P, Kreiss K, Mapp C, Milton D, Schwartz D, Toren K, Viegi G: American Thoracic Society Statement: Occupational contribution to the burden of airway disease. Am J Respir Crit Care Med 2003;167:787–797.
8   Rylander R: Endotoxin and occupational airway disease. Curr Opin Allergy Clin Immunol 2006;6:62–66.
9   Murphy TF: The role of bacteria in airway inflammation in exacerbations of chronic obstructive pulmonary disease. Curr Opin Infect Dis 2006;19:225–230.
10 Soler N, Ewig S, Torres A, Filella X, Gonzalez J, Zaubet A: Airway inflammation and bronchial microbial patterns in patients with stable chronic obstructive pulmonary disease. Eur Respir J 1999;14:1015–1022.
11 Banerjee D, Khair OA, Honeybourne D: Impact of sputum bacteria on airway inflammation and health status in clinical stable COPD. Eur Respir J 2004;23:685–691.
12 Hill AT, Campbell EJ, Hill SL, Bayley DL, Stockley RA: Association between airway bacterial load and markers of airway inflammation in patients with stable chronic bronchitis. Am J Med 2000;109:288–295.
13 Meredith TC, Aggarwal P, Mamat U, Lindner B, Woodard RW: Redefining the requisite lipopolysaccharide structure in *Escherichia coli*. ACS Chem Biol 2006;1:33–42.
14 Ernst RK, Adams KN, Moskowitz SM, Kraig GM, Kawasaki K, Stead CM, Trent MS, Miller SI: The *Pseudomonas aeruginosa* lipid A deacylase: selection for expression and loss within the cystic fibrosis airway. J Bacteriol 2006;188:191–201.

15 Montminy SW, Khan N, McGrath S, Walkowicz MJ, Sharp F, Conlon JE, Fukase K, Kusumoto S, Sweet C, Miyake K, Akira S, Cotter RJ, Goguen JD, Lien E: Virulence factors of *Yersinia pestis* are overcome by a strong lipopolysaccharide response. Nat Immunol 2006;7:1066–1073.

16 Campagnari AA, Spinola SM, Lesse AJ, Kwaik YA, Mandrell RE, Apicella MA: Lipooligosaccharide epitopes shared among gram-negative non-enteric mucosal pathogens. Microb Pathog 1990;8: 353–362.

17 Beutler B: Inferences, questions and possibilities in Toll-like receptor signalling. Nature 2004;430: 257–263.

18 Honda K, Taniguchi T: IRFs: master regulators of signalling by Toll-like receptors and cytosolic pattern-recognition receptors. Nat Rev Immunol 2006;6:644–658.

19 Pandey S, Agrawal DK: Immunobiology of Toll-like receptors: emerging trends. Immunol Cell Biol 2006;84:333–341.

20 O'Neill LA: Targeting signal transduction as a strategy to treat inflammatory diseases. Nat Rev Drug Discov 2006;5:549–563.

21 Lehner MD, Morath S, Michelsen KS, Schumann RR, Hartung T: Induction of cross-tolerance by lipopolysaccharide and highly purified lipoteichoic acid via different Toll-like receptors independent of paracrine mediators. J Immunol 2001;166:5161–5167.

22 Fahmi H, Chaby R: Desensitization of macrophages to endotoxin effects is not correlated with a down-regulation of lipopolysaccharide-binding sites. Cell Immunol 1993;150:219–229.

23 Heumann D, Gallay P, Betz-Corradin S, Barras C, Baumgartner JD, Glauser MP: Competition between bactericidal/permeability-increasing protein and lipopolysaccharide-binding protein for lipopolysaccharide binding to monocytes. J Infect Dis 1993;167:1351–1357.

24 Hasday JD, Bascom R, Costa JJ, Fitzgerald T, Dubin W: Bacterial endotoxin is an active component of cigarette smoke. Chest 1999;115:829–835.

25 Larsson L, Szponar B, Pehrson C: Tobacco smoking increases dramatically air concentrations of endotoxin. Indoor Air 2004;14:421–424.

26 Sebastian A, Pehrson C, Larsson L: Elevated concentrations of endotoxin in indoor air due to cigarette smoking. J Environ Monit 2006;8:519–522.

27 Leanord A, Williams C: *Haemophilus influenzae* in acute exacerbations of chronic obstructive pulmonary disease. Int J Antimicrob Agents 2002;19:371–375.

28 Monso E, Rosell A, Bonet G, Manterola J, Cardona PJ, Ruiz J, Morera J: Risk factors for lower airway bacterial colonization in chronic bronchitis. Eur Respir J 1999;13:338–342.

29 Cabello H, Torres A, Celis R, El-Ebiary M, Puig de la Bellacasa J, Xaubet A, Gonzalez J, Agusti C, Soler N: Bacterial colonization of distal airways in healthy subjects and chronic lung disease: a bronchoscopic study. Eur Respir J 1997;10:1137–1144.

30 Wilkinson TM, Patel IS, Wilks M, Donaldson GC, Wedzicha JA: Airway bacterial load and FEV1 decline in patients with chronic obstructive pulmonary disease. Am J Respir Crit Care Med 2003;167:1090–1095.

31 Hurst JR, Perera WR, Wilkinson TM, Donaldson GC, Wedzicha JA: Systemic and upper and lower airway inflammation at exacerbation of chronic obstructive pulmonary disease. Am J Respir Crit Care Med 2006;173:71–78.

32 Katancik JA, Kritchevsky S, Weyant RJ, Corby P, Bretz W, Crapo RO, Jensen R, Waterer G, Rubin SM, Newman AB: Periodontitis and airway obstruction. J Periodontol 2005;76:2161–2167.

33 Monso E, Ruiz J, Rosell A, Manterola J, Fiz J, Morera J, Ausina V: Bacterial infection in chronic obstructive pulmonary disease. A study of stable and exacerbated outpatients using the protected specimen brush. Am J Respir Crit Care Med 1995;152:1316–1320.

34 Soler N, Agusti C, Angrill J, Puig De la Bellacasa J, Torres A: Bronchoscopic validation of the significance of sputum purulence in severe exacerbations of chronic obstructive pulmonary disease. Thorax 2007;62:29–35.

35 Blasi F, Damato S, Cosentini R, Tarsia P, Raccanelli R, Centanni S, Allegra L: *Chlamydia pneumoniae* and chronic bronchitis: association with severity and bacterial clearance following treatment. Thorax 2002;57:672–676.

36 Nightingale JA, Rogers DF, Hart LA, Kharitonov SA, Chung KF, Barnes PJ: Effect of inhaled endotoxin on induced sputum in normal, atopic, and atopic asthmatic subjects. Thorax 1998;53: 563–571.

37   van der Veen MJ, Van Neerven RJ, De Jong EC, Aalberse RC, Jansen HM, van der Zee JS: The late asthmatic response is associated with baseline allergen-specific proliferative responsiveness of peripheral T lymphocytes in vitro and serum interleukin-5. Clin Exp Allergy 1999;29: 217–227.

38   Maris NA, de Vos AF, Dessing MC, Spek CA, Lutter R, Jansen HM, van der Zee JS, Bresser P, van der Poll T: Antiinflammatory effects of salmeterol after inhalation of lipopolysaccharide by healthy volunteers. Am J Respir Crit Care Med 2005;172:878–884.

39   Thorn J, Rylander R: Inflammatory response after inhalation of bacterial endotoxin assessed by the induced sputum technique. Thorax 1998;53:1047–1052.

40   Michel O, Dentener M, Cataldo D, Cantinieaux B, Vertongen F, Delvaux C, Murdoch RD: Evaluation of oral corticosteroids and phosphodiesterase-4 inhibitor on the acute inflammation induced by inhaled lipopolysaccharide in human. Pulm Pharmacol Ther 2006, E-pub ahead of print.

41   Michel O, Dentener M, Corazza F, Buurman W, Rylander R: Healthy subjects express differences in clinical responses to inhaled lipopolysaccharide that are related with inflammation and with atopy. J Allergy Clin Immunol 2001;107:797–804.

42   Slofstra SH, Cate HT, Spek CA: Low dose endotoxin priming is accountable for coagulation abnormalities and organ damage observed in the Shwartzman reaction. A comparison between a single-dose endotoxemia model and a double-hit endotoxin-induced Shwartzman reaction. Thromb J 2006;4:13.

43   O'Grady NP, Preas HL, Pugin J, Fiuza C, Tropea M, Reda D, Banks SM, Suffredini AF: Local inflammatory responses following bronchial endotoxin instillation in humans. Am J Respir Crit Care Med 2001;163:1591–1598.

44   Kharitonov SA, Barnes PJ: Exhaled biomarkers. Chest 2006;130:1541–1546.

45   Crapo RO, Jensen RL, Hargreave FE: Airway inflammation in COPD: physiological outcome measures and induced sputum. Eur Respir J Suppl 2003;41:19s–28s.

46   Brindicci C, Ito K, Resta O, Pride NB, Barnes PJ, Kharitonov SA: Exhaled nitric oxide from lung periphery is increased in COPD. Eur Respir J 2005;26:52–59.

47   Rolla G, Bucca C, Brussino L, Dutto L, Colagrande P, Polizzi S: Pentoxifylline attenuates LPS-induced bronchial hyperresponsiveness but not the increase in exhaled nitric oxide. Clin Exp Allergy 1997;27:96–103.

48   Paredi P, Kharitonov SA, Barnes PJ: Correlation of exhaled breath temperature with bronchial blood flow in asthma. Respir Res 2005;6:15.

49   Paredi P, Caramori G, Cramer D, Ward S, Ciaccia A, Papi A, Kharitonov SA, Barnes PJ: Slower rise of exhaled breath temperature in chronic obstructive pulmonary disease. Eur Respir J 2003;21:439–443.

50   Kharitonov SA, Barnes PJ: Clinical aspects of exhaled nitric oxide. Eur Respir J 2000;16: 781–792.

51   Garey KW, Neuhauser MM, Robbins RA, Danziger LH, Rubinstein I: Markers of inflammation in exhaled breath condensate of young healthy smokers. Chest 2004;125:22–26.

52   Griese M, Noss J, von Bredow C: Protein pattern of exhaled breath condensate and saliva. Proteomics 2002;2:690–696.

53   Alexis N, Eldridge M, Reed W, Bromberg P, Peden DB: CD14-dependent airway neutrophil response to inhaled LPS: role of atopy. J Allergy Clin Immunol 2001;107:31–35.

54   Wallin A, Pourazar J, Sandstrom T: LPS-induced bronchoalveolar neutrophilia; effects of salmeterol treatment. Respir Med 2004;98:1087–1092.

55   Maris NA, Florquin S, van't Veer C, de Vos AF, Buurman W, Jansen HM, van der Poll T: Inhalation of beta 2 agonists impairs the clearance of nontypable *Haemophilus influenzae* from the murine respiratory tract. Respir Res 2006;7:57.

56   Xiao W, Hsu YP, Ishizaka A, Kirikae T, Moss RB: Sputum cathelicidin, urokinase plasminogen activation system components, and cytokines discriminate cystic fibrosis, COPD, and asthma inflammation. Chest 2005;128:2316–2326.

57   Krug NLM, Bonner J, Mueller M, Erenbeck VJ, Schauman: Anti-MCP-1 monoclonal antibody (ABN912) attenuates lipopolysaccharide-induced neutrophilic inflammation. Am J Respir Crit Care Med 2006;A849.

58   Hong MLIK, Donnelly L, Barnes PJ, Kharitonov SA: Effect of endotoxin on inflammatory markers in exhaled breath, sputum and nasal lavage, non-smokers and current smokers. Am J Respir Crit Care Med 2004;A76.

59   Di Stefano A, Caramori G, Oates T, Capelli A, Lusuardi M, Gnemmi I, Ioli F, Chung KF, Donner CF, Barnes PJ, Adcock IM: Increased expression of nuclear factor-kappaB in bronchial biopsies from smokers and patients with COPD. Eur Respir J 2002;20:556–563.

60   Maris NA, de Vos AF, Bresser P, van der Zee JS, Jansen HM, Levi M, van der Poll T: Salmeterol enhances pulmonary fibrinolysis in healthy volunteers. Crit Care Med 2007;35:57–63.

61   Spond J, Chapman R, Fine J, Jones H, Kreutner W, Kung TT, Minnicozzi M: Comparison of PDE 4 inhibitors, rolipram and SB 207499 (ariflo), in a rat model of pulmonary neutrophilia. Pulm Pharmacol Ther 2001;14:157–164.

62   Hurst JR, Wilkinson TM, Perera WR, Donaldson GC, Wedzicha JA: Relationships among bacteria, upper airway, lower airway, and systemic inflammation in COPD. Chest 2005;127:1219–1226.

63   Roberts NJ, Lloyd-Owen SJ, Rapado F, Patel IS, Wilkinson TM, Donaldson GC, Wedzicha JA: Relationship between chronic nasal and respiratory symptoms in patients with COPD. Respir Med 2003;97:909–914.

64   Keman S, Jetten M, Douwes J, Borm PJ: Longitudinal changes in inflammatory markers in nasal lavage of cotton workers. Relation to endotoxin exposure and lung function changes. Int Arch Occup Environ Health 1998;71:131–137.

65   Lowry SF: Human endotoxemia: a model for mechanistic insight and therapeutic targeting. Shock 2005;24(suppl 1):94–100.

66   Suffredini AF, Harpel PC, Parrillo JE: Promotion and subsequent inhibition of plasminogen activation after administration of intravenous endotoxin to normal subjects. N Engl J Med 1989;320:1165–1172.

67   Preas HL 2nd, Jubran A, Vandivier RW, Reda D, Godin PJ, Banks SM, Tobin MJ, Suffredini AF: Effect of endotoxin on ventilation and breath variability: role of cyclooxygenase pathway. Am J Respir Crit Care Med 2001;164:620–626.

68   Lynn M, Rossignol DP, Wheeler JL, Kao RJ, Perdomo CA, Noveck R, Vargas R, D'Angelo T, Gotzkowsky S, McMahon FG: Blocking of responses to endotoxin by E5564 in healthy volunteers with experimental endotoxemia. J Infect Dis 2003;187:631–639.

Dr. Sergei A. Kharitonov, MD, PhD
Section of Airway Disease, National Heart and Lung Institute, Imperial College London
Dovehouse Street
London SW3 6LY (UK)
Tel. +44 020 7351 8006, Fax +44 020 7351 8126, E-Mail s.kharitonov@imperial.ac.uk

Sjöbring U, Taylor JD (eds): Models of Exacerbations in Asthma and COPD.
Contrib Microbiol. Basel, Karger, 2007, vol 14, pp 101–112

A Human Rhinovirus Model of Chronic Obstructive Pulmonary Disease Exacerbations

Marco Contoli[a,b], *Gaetano Caramori*[a], *Patrick Mallia*[b], *Alberto Papi*[a],
Sebastian L. Johnston[b]

[a]Department of Clinical and Experimental Medicine, Research Center on Asthma and
COPD, University of Ferrara, Ferrara, Italy; [b]Department of Respiratory Medicine,
National Heart and Lung Institute and Wright Fleming Institute of Infection and
Immunity, Imperial College London, London, UK

Abstract

Chronic obstructive pulmonary disease (COPD) exacerbations are common events that
punctuate the natural history of COPD contributing to disease severity progression and being the
major cause of COPD-related morbidity and mortality. Currently available pharmacological
strategies are only partially effective at reducing or preventing COPD exacerbations. Viral infections are the most frequent cause of COPD exacerbations. The recent development of a human
experimental model of rhinovirus-induced COPD exacerbations represents an innovative tool
with the potential to increase our understanding of the inflammatory and immunological mechanisms that lead COPD patients to exacerbate after respiratory virus infections. Moreover this
model will provide the opportunity to test, in a carefully controlled setting, novel pharmacological compounds with a potential for treating and preventing COPD exacerbations. In this chapter
we will focus on the role of viral infections in COPD exacerbations and will discuss preliminary
reports regarding the development of this human model of virus-induced COPD exacerbation.

Introduction

The clinical history of chronic obstructive pulmonary disease (COPD) is
characterized by recurrent episodes of increased in dyspnoea, cough or sputum
production sufficient to warrant a change in management. These events are
named exacerbations and are a common occurrence in many COPD patients.
The frequency of COPD exacerbations increases with increased severity of the

disease [1–3]. In addition to increasing COPD-associated morbidity and mortality, exacerbations contribute to loss of lung function and impaired health status in COPD patients [4–7]. Thus treatment and prevention of COPD exacerbations are fundamental to the management of COPD. Nevertheless the available pharmacological treatment of COPD is only partially effective in preventing COPD exacerbations.

COPD exacerbations are caused or triggered by a variety of factors including bacteria, viruses, air pollutants and changes of temperature and are associated with acute worsening of the pre-existing lower airway inflammation. In the last decade the development of highly sensitive diagnostic tools has provided strong evidence that respiratory viral infections are one of the major causes of COPD exacerbations. Nevertheless, the immunological and inflammatory mechanisms that lead to exacerbation in COPD patients after respiratory viral infections are almost unknown.

The development of a human in vivo model of virus-induced COPD exacerbation is a fundamental step towards the knowledge of these mechanisms providing an invaluable tool to highlight and to test novel pharmacological targets for treating and preventing virus-induced COPD exacerbations.

Virus Infection in COPD

COPD exacerbations are associated with an increased number of activated inflammatory cells in the sputum and a changed pattern of proinflammatory and anti-inflammatory mediators released into this compartment [7–9]. Through influencing the inflammatory response respiratory viral infections may enhance the pathological processes associated with cigarette smoking and contribute to the lung pathology and loss of lung function associated with COPD. The type and the degree of activation of the different inflammatory cells recruited to the lung during COPD exacerbations have not been well characterized.

CD8+ T lymphocytes, infiltrating both central [10] and peripheral airways [11], are a hallmark of COPD inflammation. These cells are further increased in the sputum during exacerbations [12]. Among the possible causes of CD8+ recruitment in the airways, chronic viral infection of the lungs has been suggested to be important. Particular attention has been given to the possible role of latent adenoviral infection in the pathogenesis of stable COPD. Adenoviral DNA sequences encoding the E1A protein are increased in the lung tissue of COPD patients as compared to smokers with normal lung function, suggesting that latent adenoviral infection may occur in the lungs of COPD patients [13]. Moreover, in COPD patients with emphysema the presence of adenoviral E1A protein in the alveolar epithelial cells was related to the severity of the

emphysematous lesions and the number of CD8+ cells infiltrating the lung
[14]. These observations are consistent with the hypothesis that a persistent
intracellular pathogen such as adenovirus may be capable of amplifying cigarette smoke-induced inflammation, possibly through interaction of viral proteins (such as E1A) with proinflammatory transcription factors [e.g. nuclear
factor-κB (NF-κB) and activator protein-1 (AP-1)] [14]. Recently it has also
been documented that COPD patients in whom respiratory syncytial virus was
repeatedly detected in sputum over 2 years had faster lung function decline
[15]. Taken together these data indicate that in response to repeated/latent viral
infections, an excessive recruitment of CD8+ T lymphocytes may occur in the
tracheobronchial tree. The CD8+ driven inflammation can damage the lung in
susceptible smokers leading to COPD progression [16].

The use of highly sensitive diagnostic methods such as polymerase chain
reaction (PCR) to evaluate the association between respiratory virus infections
and COPD exacerbations has shown that viruses are responsible for a much
higher proportion of exacerbations than was previously realized. In a study of
the East London COPD cohort, PCR was used to detect rhinovirus in nasal and
sputum samples from COPD patients during an exacerbation and when clinically stable [17]. Twenty-three percent of exacerbation samples were positive
for rhinovirus compared to 0% when clinically stable. These data clearly indicate that, at variance with bacteria in which a chronic colonization can be present also in the stable state, rhinovirus infection is important in inducing
exacerbation. In a further study virus was detected in 39% of exacerbations, the
most common being rhinoviruses that accounted for 58% of viruses [18].

A high virus detection rate has also been reported in hospitalized patients.
A respiratory virus was detected in around 50% of patients with severe COPD
exacerbation admitted to hospitals in Germany and in Italy, with rhinovirus
again being the most common [19, 20]. In patients with COPD requiring intu-
bation and mechanical ventilation virus was identified in 47% of patients [21].
Together, these data indicate that respiratory virus infection is associated with a
substantial proportion of COPD exacerbations, with rhinovirus being the most
frequently identified virus.

A recent study has addressed the relative importance of viral versus bacterial infections to the aetiology of severe (hospitalized) COPD exacerbations.
Viral and/or bacterial infection was detected in 78% of COPD exacerbations,
with viruses in 48.4% (6.2% when stable) and bacteria in 54.7% (37.5% when
stable). Patients with exacerbations of infectious aetiology required longer hospitalizations and demonstrated a greater impairment of several measures of
lung function than patients with non-infectious exacerbations. Moreover, the
most severe exacerbations were those in which viral and bacterial co-infection
was detected [20]. Similar results also have been found in studies of COPD

exacerbations in outpatients [22]. Indeed patients, in whom both bacterial and viral pathogens were detected, had increased inflammatory markers in sputum and a greater lung function fall as compared to patients in whom a single pathogen was detected. However, the relationship between viral and bacterial infection, especially when combined, needs to be further studied. In particular it needs to be established whether viral infection can pave the way for exacerbation by bacteria colonizing the lower respiratory tract of COPD patients.

Mechanisms of Virus-Induced COPD Exacerbations

At variance with asthma in which several in vitro and in vivo studies have investigated the mechanisms that lead to exacerbation after viral infection of the airways [23, 24], very few data are available for COPD.

Inflammatory Mechanisms
During COPD exacerbations several inflammatory mediators are increased in the airways over baseline levels. Moreover COPD patients with more frequent exacerbations have higher sputum levels of the inflammatory markers interleukin-6 (IL-6) and IL-8 even when stable [25], suggesting that there may be proinflammatory effects of exacerbations that persist after the acute episode has resolved. Little is known about the inflammatory mediators specific to virus-induced COPD exacerbations. One study has found an increased sputum level of IL-6 in viral-associated acute episodes as compared to non-viral exacerbations [17]. Recently the simultaneous presence of both rhinovirus and *Haemophilus influenzae* at exacerbations has been associated with an increased level of serum IL-6 suggesting that viruses and bacteria can synergistically interact to increase the severity of the inflammation [22].

The major group of rhinoviruses (accounting for 90% of total rhinovirus types) attaches to airway epithelium by binding to the intercellular adhesion molecule 1 (ICAM-1) [26]. Interestingly, rhinovirus infection induces expression of its own receptor ICAM-1 [27, 28], which may promote inflammatory cell recruitment and activation. There is evidence for upregulation of ICAM-1 in the bronchial mucosa of patients with chronic bronchitis [29]. Upregulation of ICAM-1 may increase infectivity and lead to an augmented inflammatory response.

Cellular Mechanisms
The type of inflammatory cells recruited to the lung during COPD exacerbations has not been fully clarified. Very few studies have analyzed bronchial biopsies at exacerbations due to the difficulties of carrying out an invasive procedure in acutely ill patients. Two studies using bronchial biopsies during

exacerbations of chronic bronchitis from a single cohort of patients reported a prominent airway eosinophilia at exacerbation together with an increased number of neutrophils and T lymphocytes in the exacerbated group compared to the stable patients [30, 31]. A recent study has shown an increased number of neutrophils in sputum during exacerbations and the increase was independent of the type (bacteria or virus) of infectious agent detected. The same study documented that virus-induced COPD exacerbations with or without concomitant bacterial infection are associated with an increased number of sputum eosinophils, suggesting that sputum eosinophilia could be a marker of viral infection during COPD exacerbations [20].

Interestingly, increased sputum CD8+ T lymphocytes have been reported during COPD exacerbations with a relative reduction in the ratio of interferon-γ (IFN-γ)/IL-4 expressing CD8+ T lymphocytes [12]. Thus, a switch towards a T helper 2 (Th2)-like immunophenotype during COPD exacerbations could trigger recruitment of eosinophils, a classical effector cell recruited during Th2-mediated immune responses, during virus-induced COPD exacerbations.

Oxidative Stress and Proinflammatory Intracellular Signalling Pathways

COPD is associated with increased oxidative stress that is believed to play a central role in the disease pathogenesis. Oxidants could represent a key intracellular mediator of virus-induced cellular activation. Interestingly, in vitro oxidative stress can induce activation of NF-κB and AP-1, two pivotal regulators of the inflammatory processes. NF-κB is activated in sputum macrophages during COPD exacerbations [32]. Rhinovirus infection induces increased production of superoxide anion in bronchial epithelial cells and this event is a crucial step for the activation of NF-κB and the following production of proinflammatory cytokines, chemokines and adhesion molecules [33]. Reducing agents inhibit both rhinovirus-induced oxidant generation and inflammatory mediator production and release. These data suggest that the inhibition of intracellular oxidative stress may be a potential therapeutic target for the treatment of virus-induced COPD exacerbations [33].

Susceptibility to Virus Infections in COPD Patients

There is solid evidence of impaired innate [34, 35] and possibly acquired [36, 37] immune responses to viral infection in asthmatic patients. Whether COPD patients are more susceptible to virus infection as compared to normal subjects is still debated. A recent study documented that patients with frequent (>2.5/year) COPD exacerbation have more frequent episodes of naturally occurring colds as compared to patients with infrequent exacerbations [38]. These results suggest that COPD subjects with frequent exacerbations may represent a subgroup particularly susceptible to viral infections, but they do not determine whether this susceptibility relates to the general COPD population. Intriguingly patients experiencing frequent colds had a significantly higher exposure to cigarette smoke [38]. Recently, using a murine model of cigarette smoke exposure, it was demonstrated that virus infections in combination with tobacco smoke decreased the number of dendritic cells in the lung and also altered the costimulatory molecule expression profile on these cells that are believed to be fundamental to the initiation of adaptive immune response [39]. In the same study it was also found that mice exposed to smoke after adenovirus infection had a decreased T cell-mediated antiviral immune response, as documented by reduced infiltration of activated CD4 and CD8 T cells, in the lungs [39]. Thus cigarette smoke, possibly via alteration of both innate and adaptive immune responses, can contribute to the increased susceptibility of COPD patients to viral infections [40].

Another possible mechanism leading to increased susceptibility is related to upregulation of ICAM-1, the receptor for the major group of human rhinoviruses. Latent expression of adenoviral E1A protein in alveolar epithelial cells of patients with pulmonary emphysema may increase ICAM-1 expression and this could be a potential mechanism for a greater susceptibility to rhinovirus infection of COPD patients [14].

Finally, solid evidence exists that the surface of the bronchial mucosa of patients with stable COPD is chronically colonized with bacteria, and the bacterial load is related to the intensity of the airway inflammation and to disease progression [41]. Moreover patients with a history of frequent exacerbations have a higher incidence of bacterial colonization. Thus, given that viral infections are one of the most frequent causes of COPD exacerbation these results indicate that chronic bacterial colonization could contribute to the increased susceptibility of COPD patients to viral infection, for example by increasing ICAM-1 expression on the surface of the bronchial epithelial cells [42]. Further studies are required to investigate the interaction between chronic bacterial colonization and respiratory viral infection and in particular whether chronic bacterial colonization can increase viral infection susceptibility or alternatively whether viral infection can increase the airway bacteria load leading to COPD exacerbations.

In vivo Model of Viral-Induced COPD Exacerbation

In vitro data document that respiratory virus infection can lead to COPD exacerbation via the production of several proinflammatory molecules that are relevant to the pathogenesis of COPD exacerbation [28, 43–45]. However, although in vitro models can provide important insights into the molecular mechanisms of inflammatory and immune responses to viral and bacterial infections, the in vitro data require validation using in vivo models.

Bacterial infections have been considered important causes of COPD exacerbations for a long time. Nevertheless, while several mechanisms (e.g. induction of mucus hypersecretion [46], reduction of ciliary beat frequency [47] and enhancement of neutrophilic inflammation [48]) have been proposed to explain how bacterial infection can trigger COPD exacerbations, no animal model of bacteria-induced COPD exacerbation is available. Until recently this was also true for viral exacerbations.

Carrying out studies of naturally occurring COPD exacerbations has proved difficult for a number of reasons including non-reporting of exacerbations by patients, lack of baseline data before exacerbations, wide variation in aetiology, variation in timing of sampling relative to onset of exacerbation and finally carrying out invasive airway investigations in acutely unwell patients, which is difficult and may jeopardize their health. One way to overcome these obstacles is the development of a human experimental model that would allow studies to take place under controlled conditions. The first step towards the development of such a model has recently been realized [49] with the reporting of the first study evaluating the effects of an experimental rhinovirus infection in COPD patients.

In this pilot study mild to moderate COPD patients were selected for experimental infection with the purpose (1) of evaluating whether the procedure is safe in patients with mild to moderate COPD (mean FEV_1 was 74.8% predicted) and (2) of providing preliminary data on whether experimental rhinovirus infection in COPD patients is per se sufficient to trigger an exacerbation. As safety was the prime concern, a very carefully designed rhinovirus dose escalation study was performed in a small (n = 5) group of subjects to determine the minimum dose of virus able to induce clinical colds in 80% (4 out of 5) of the inoculated subjects. Surprisingly, all of the first 4 patients exposed to the initial lowest dose of rhinovirus inoculum experienced not only cold symptoms but also lower respiratory tract symptoms characteristic of a COPD exacerbation associated with a significant fall in lung function [49] as occurs with naturally occurring exacerbations [50]. The severity of the exacerbations induced was graded mild to moderate, thus achieving the primary aim of the study to show that experimental rhinovirus infection could be safely carried out in mild to moderate COPD patients.

This model also showed that experimental rhinovirus infection can cause exacerbation in COPD patients and has the potential to provide a valid model of naturally occurring COPD exacerbation. In COPD this is so far the only scenario in which a specific aetiology has been experimentally proven to induce exacerbation.

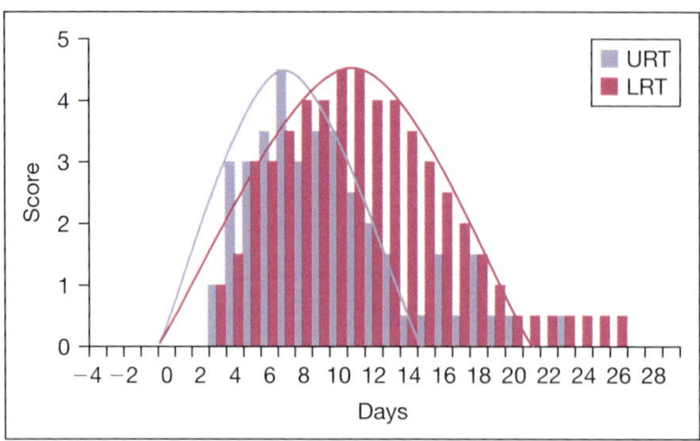

Fig. 1. Upper (URT) and lower (LRT) respiratory tract symptoms following an in vivo rhinovirus experimental infection in COPD patients. A 3- to 4-day gap between the peak of cold symptoms and the peak of lower respiratory symptoms was documented [49]. This data suggests that if an effective antiviral or anti-inflammatory treatment could be given at the onset of cold symptoms, this could possibly change the clinical outcome of the viral infection in COPD.

Two potentially important preliminary observations arose from this study: (1) COPD patients developed colds and exacerbations with 100- to 1,000-fold lower doses of virus than used in previous studies in asthmatic and normal volunteers, and (2) there was a 3- to 4-day gap between the peak of cold symptoms and the peak of lower respiratory symptoms [49] (fig. 1). These data suggest that COPD patients may be highly susceptible to virus infection and that if an effective antiviral or anti-inflammatory treatment could be given at the onset of cold symptoms, this could possibly change the clinical outcome of the viral infection in COPD. These preliminary findings clearly require confirmation in a larger study of COPD patients.

In a similar experimental model it has recently been documented in asthmatic patients that the severity of the virus-induced exacerbations, both in terms of symptoms, lung function reduction and inflammation in the airways, was inversely related to the production of a novel class of interferons called IFN-λ [35]. An impaired innate immune response can therefore be one of the mechanisms of increased susceptibility of asthmatic patients to respiratory viral infections. The new model in COPD will give the opportunity to test whether similar or different immune deficiencies are present in COPD, and if so, whether modulation and/or restoring of the immune response could represent a plausible pharmacological approach to treat/prevent COPD exacerbations.

To further validate the results achieved in the in vivo experimental model in COPD and to investigate more deeply the mechanisms of viral-induced COPD exacerbations, a larger study has recently been undertaken and preliminary findings published in form of an abstract [51]. In this study 21 subjects (10 mild COPD patients and 11 age- and smoking-matched controls) have been enrolled. The results confirm that the model is feasible and safe. Symptomatic colds were accompanied by lower respiratory tract symptoms in both groups with breathlessness increasing significantly only in the COPD group. Interestingly, at variance with smokers with normal lung function, sputum neutrophils significantly increased in COPD patients following experimental infection. These data indicate that experimental rhinovirus infection in COPD induces symptoms, lung function changes and systemic and airway inflammation similar to that observed in naturally occurring exacerbations supporting, for the first time, a causal relationship between rhinovirus infection and COPD exacerbations. The development of such an experimental model, in which causation is clearly defined and in which detailed clinical studies on mechanisms of disease can be carried out, will offer an invaluable tool to increase our understanding of the specific immunological and inflammatory events that lead COPD patients to exacerbate after viral infection. Moreover this model will offer the possibility to highlight and to test novel pharmacological targets able to treat and/or prevent viral-induced COPD exacerbations.

Conclusions

COPD is a major health problem worldwide with rising prevalence and mortality. The major morbidity, mortality and health care costs of COPD are due to exacerbations [52].

Thanks to the development of highly sensitive diagnostic tools, respiratory viruses have emerged as a leading cause of COPD exacerbations. The mechanisms responsible for respiratory virus-induced COPD exacerbations are still largely unknown.

The recent development of the first human model of virus-induced COPD exacerbation [49] will facilitate identification of novel pharmacological targets that will provide opportunities to develop new treatments for COPD exacerbations.

References

1 Donaldson GC, Seemungal TA, Patel IS, Lloyd-Owen SJ, Wilkinson TM, Wedzicha JA: Longitudinal changes in the nature, severity and frequency of COPD exacerbations. Eur Respir J 2003;22:931–936.
2 Decramer M, Rutten-van Molken MP, Dekhuijzen PN, Troosters T, van Herwaarden C, Pellegrino R, van Schayck CP, Olivieri D, Del Donno M, De Backer W, Lankhorst I, Ardia A: Effects of N-acetylcysteine on outcomes in chronic obstructive pulmonary disease (Bronchitis Randomized

on NAC Cost-Utility Study, BRONCUS): a randomized placebo-controlled trial. Lancet 2005;365: 1552–1560.

3 Anthonisen NR, Manfreda J, Warren CP, Hershfield ES, Harding GK, Nelson NA: Antibiotic therapy in exacerbations of chronic obstructive pulmonary disease. Ann Intern Med 1987;106: 196–204.

4 Donaldson GC, Seemungal TA, Bhowmik A, Wedzicha JA: Relationship between exacerbation frequency and lung function decline in chronic obstructive pulmonary disease. Thorax 2002;57: 847–852.

5 Kanner RE, Anthonisen NR, Connett JE: Lower respiratory illnesses promote FEV(1) decline in current smokers but not ex-smokers with mild chronic obstructive pulmonary disease: results from the lung health study. Am J Respir Crit Care Med 2001;164:358–364.

6 Seemungal TA, Donaldson GC, Paul EA, Bestall JC, Jeffries DJ, Wedzicha JA: Effect of exacerbation on quality of life in patients with chronic obstructive pulmonary disease. Am J Respir Crit Care Med 1998;157:1418–1422.

7 Papi A, Luppi F, Franco F, Fabbri LM: Pathophysiology of exacerbations of chronic obstructive pulmonary disease. Proc Am Thorac Soc 2006;3:245–251.

8 Caramori G, Adcock I: Pharmacology of airway inflammation in asthma and COPD. Pulm Pharmacol Ther 2003;16:247–277.

9 Wedzicha JA, Donaldson GC: Exacerbations of chronic obstructive pulmonary disease. Respir Care 2003;48:1204–1213.

10 O'Shaughnessy TC, Ansari TW, Barnes NC, Jeffery PK: Inflammation in bronchial biopsies of subjects with chronic bronchitis: inverse relationship of CD8+ T lymphocytes with FEV1. Am J Respir Crit Care Med 1997;155:852–857.

11 Saetta M, Di Stefano A, Turato G, Facchini FM, Corbino L, Mapp CE, Maestrelli P, Ciaccia A, Fabbri LM: CD8+ T-lymphocytes in peripheral airways of smokers with chronic obstructive pulmonary disease. Am J Respir Crit Care Med 1998;157:822–826.

12 Tsoumakidou M, Tzanakis N, Chrysofakis G, Kyriakou D, Siafakas NM: Changes in sputum T-lymphocyte subpopulations at the onset of severe exacerbations of chronic obstructive pulmonary disease. Respir Med 2005;99:572–579.

13 Matsuse T, Hayashi S, Kuwano K, Keunecke H, Jefferies WA, Hogg JC: Latent adenoviral infection in the pathogenesis of chronic airways obstruction. Am Rev Respir Dis 1992;146:177–184.

14 Retamales I, Elliott WM, Meshi B, Coxson HO, Pare PD, Sciurba FC, Rogers RM, Hayashi S, Hogg JC: Amplification of inflammation in emphysema and its association with latent adenoviral infection. Am J Respir Crit Care Med 2001;164:469–473.

15 Wilkinson TM, Donaldson GC, Johnston SL, Openshaw PJ, Wedzicha JA: Respiratory syncytial virus, airway inflammation, and FEV1 decline in patients with chronic obstructive pulmonary disease. Am J Respir Crit Care Med 2006;173:871–876.

16 Majo J, Ghezzo H, Cosio MG: Lymphocyte population and apoptosis in the lungs of smokers and their relation to emphysema. Eur Respir J 2001;17:946–953.

17 Seemungal TA, Harper-Owen R, Bhowmik A, Jeffries DJ, Wedzicha JA: Detection of rhinovirus in induced sputum at exacerbation of chronic obstructive pulmonary disease. Eur Respir J 2000;16: 677–683.

18 Seemungal T, Harper-Owen R, Bhowmik A, Moric I, Sanderson G, Message S, Maccallum P, Meade TW, Jeffries DJ, Johnston SL, Wedzicha JA: Respiratory viruses, symptoms, and inflammatory markers in acute exacerbations and stable chronic obstructive pulmonary disease. Am J Respir Crit Care Med 2001;164:1618–1623.

19 Rohde G, Wiethege A, Borg I, Kauth M, Bauer TT, Gillissen A, Bufe A, Schultze-Werninghaus G: Respiratory viruses in exacerbations of chronic obstructive pulmonary disease requiring hospitalisation: a case-control study. Thorax 2003;58:37–42.

20 Papi A, Bellettato CM, Braccioni F, Romagnoli M, Casolari P, Caramori G, Fabbri LM, Johnston SL: Infections and airway inflammation in chronic obstructive pulmonary disease severe exacerbations. Am J Respir Crit Care Med 2006;173:1114–1121.

21 Qiu Y, Zhu J, Bandi V, Atmar RL, Hattotuwa K, Guntupalli KK, Jeffery PK: Biopsy neutrophilia, neutrophil chemokine and receptor gene expression in severe exacerbations of chronic obstructive pulmonary disease. Am J Respir Crit Care Med 2003;168:968–975.

22 Wilkinson TM, Hurst JR, Perera WR, Wilks M, Donaldson GC, Wedzicha JA: Effect of interactions between lower airway bacterial and rhinoviral infection in exacerbations of COPD. Chest 2006;129:317–324.

23 Caramori G, Ito K, Contoli M, Di Stefano A, Johnston SL, Adcock IM, Papi A: Molecular mechanisms of respiratory virus-induced asthma and COPD exacerbations and pneumonia. Curr Med Chem 2006;13:2267–2290.

24 Contoli M, Caramori G, Mallia P, Johnston S, Papi A: Mechanisms of respiratory virus-induced asthma exacerbations. Clin Exp Allergy 2005;35:137–145.

25 Bhowmik A, Seemungal TA, Sapsford RJ, Wedzicha JA: Relation of sputum inflammatory markers to symptoms and lung function changes in COPD exacerbations. Thorax 2000;55:114–120.

26 Greve JM, Davis G, Meyer AM, Forte CP, Yost SC, Marlor CW, Kamarck ME, McClelland A: The major human rhinovirus receptor is ICAM-1. Cell 1989;56:39–47.

27 Grunberg K, Sharon RF, Hiltermann TJ, Brahim JJ, Dick EC, Sterk PJ, Van Krieken JH: Experimental rhinovirus 16 infection increases intercellular adhesion molecule-1 expression in bronchial epithelium of asthmatics regardless of inhaled steroid treatment. Clin Exp Allergy 2000;30:1015–1023.

28 Papi A, Johnston SL: Rhinovirus infection induces expression of its own receptor intercellular adhesion molecule 1 (ICAM-1) via increased NF-kappaB-mediated transcription. J Biol Chem 1999;274:9707–9720.

29 Vignola AM, Campbell AM, Chanez P, Bousquet J, Paul-Lacoste P, Michel FB, Godard P: HLA-DR and ICAM-1 expression on bronchial epithelial cells in asthma and chronic bronchitis. Am Rev Respir Dis 1993;148:689–694.

30 Saetta M, Di Stefano A, Maestrelli P, Turato G, Ruggieri MP, Roggeri A, Calcagni P, Mapp CE, Ciaccia A, Fabbri LM: Airway eosinophilia in chronic bronchitis during exacerbations. Am J Respir Crit Care Med 1994;150:1646–1652.

31 Zhu J, Qiu YS, Majumdar S, Gamble E, Matin D, Turato G, Fabbri LM, Barnes N, Saetta M, Jeffery PK: Exacerbations of bronchitis: bronchial eosinophilia and gene expression for interleukin-4, interleukin-5, and eosinophil chemoattractants. Am J Respir Crit Care Med 2001;164: 109–116.

32 Caramori G, Romagnoli M, Casolari P, Bellettato C, Casoni G, Boschetto P, Chung KF, Barnes PJ, Adcock IM, Ciaccia A, Fabbri LM, Papi A: Nuclear localisation of p65 in sputum macrophages but not in sputum neutrophils during COPD exacerbations. Thorax 2003;58:348–351.

33 Papi A, Papadopoulos NG, Stanciu LA, Bellettato CM, Pinamonti S, Degitz K, Holgate ST, Johnston SL: Reducing agents inhibit rhinovirus-induced up-regulation of the rhinovirus receptor intercellular adhesion molecule-1 (ICAM-1) in respiratory epithelial cells. FASEB J 2002;16: 1934–1936.

34 Wark P, Johnston SL, Bucchieri F, Powell R, Puddicombe S, Laza-Stanca V, Holgate ST, Davies DE: Asthmatic bronchial epithelial cells have a deficient innate immune response to infection with rhinovirus. J Exp Med 2005;201:937–947.

35 Contoli M, Message SD, Laza-Stanca V, Edwards MR, Wark PA, Bartlett NW, Kebadze T, Mallia P, Stanciu LA, Parker HL, Slater L, Lewis-Antes A, Kon OM, Holgate ST, Davies DE, Kotenko SV, Papi A, Johnston SL: Role of deficient type III interferon-lambda production in asthma exacerbations. Nat Med 2006;12:1023–1026.

36 Papadopoulos NG, Stanciu LA, Papi A, Holgate ST, Johnston SL: A defective type 1 response to rhinovirus in atopic asthma. Thorax 2002;57:328–332.

37 Gern JE, Vrtis R, Grindle KA, Swenson C, Busse WW: Relationship of upper and lower airway cytokines to outcome of experimental rhinovirus infection. Am J Respir Crit Care Med 2000; 162:2226–2231.

38 Hurst JR, Donaldson GC, Wilkinson TM, Perera WR, Wedzicha JA: Epidemiological relationships between the common cold and exacerbation frequency in COPD. Eur Respir J 2005;26: 846–852.

39 Robbins CS, Dawe DE, Goncharova SI, Pouladi MA, Drannik AG, Swirski FK, Cox G, Stampfli MR: Cigarette smoke decreases pulmonary dendritic cells and impacts antiviral immune responsiveness. Am J Respir Cell Mol Biol 2004;30:202–211.

40 Sopori M: Effects of cigarette smoke on the immune system. Nat Rev Immunol 2002;2:372–377.

41 Patel IS, Seemungal TA, Wilks M, Lloyd-Owen SJ, Donaldson GC, Wedzicha JA: Relationship between bacterial colonisation and the frequency, character, and severity of COPD exacerbations. Thorax 2002;57:759–764.
42 Sajjan US, Jia Y, Newcomb DC, Bentley JK, Lukacs NW, LiPuma JJ, Hershenson MB: *H. influenzae* potentiates airway epithelial cell responses to rhinovirus by increasing ICAM-1 and TLR3 expression. FASEB J 2006;20:2121–2123.
43 Donninger H, Glashoff R, Haitchi HM, Syce JA, Ghildyal R, van Rensburg E, Bardin PG: Rhinovirus induction of the CXC chemokine epithelial-neutrophil activating peptide-78 in bronchial epithelium. J Infect Dis 2003;187:809–817.
44 Johnston SL, Papi A, Bates PJ, Mastronarde JG, Monick MM, Hunninghake GW: Low grade rhinovirus infection induces a prolonged release of IL-8 in pulmonary epithelium. J Immunol 1998;160:6172–6181.
45 Laza-Stanca V, Stanciu LA, Message SD, Edwards MR, Gern JE, Johnston SL: Rhinovirus replication in human macrophages induces NF-kappaB-dependent tumor necrosis factor alpha production. J Virol 2006;80:8248–8258.
46 Adler KB, Hendley DD, Davis GS: Bacteria associated with obstructive pulmonary disease elaborate extracellular products that stimulate mucin secretion by explants of guinea pig airways. Am J Pathol 1986;125:501–514.
47 Wilson R, Roberts D, Cole P: Effect of bacterial products on human ciliary function in vitro. Thorax 1985;40:125–131.
48 Read RC, Wilson R, Rutman A, Lund V, Todd HC, Brain AP, Jeffery PK, Cole PJ: Interaction of nontypable *Haemophilus influenzae* with human respiratory mucosa in vitro. J Infect Dis 1991;163: 549–558.
49 Mallia P, Message SD, Kebadze T, Parker HL, Kon OM, Johnston SL: An experimental model of rhinovirus induced chronic obstructive pulmonary disease exacerbations: a pilot study. Respir Res 2006;7:116.
50 Seemungal TA, Donaldson GC, Bhowmik A, Jeffries DJ, Wedzicha JA: Time course and recovery of exacerbations in patients with chronic obstructive pulmonary disease. Am J Respir Crit Care Med 2000;161:1608–1613.
51 Mallia P, Message S, Contoli M, Gray K, Kebadze T, Laza-Stanca V, Kon O, Johnston SL: An experimental model of virus-induced COPD exacerbation. Thorax 2006;61:9(suppl):Abstract S076.
52 NHLBI/WHO Workshop report. Global Initiative for Chronic Obstructive Lung Disease (GOLD): Global strategy for the diagnosis, management and prevention of chronic obstructive pulmonary disease. NIH Publication No 2701A, March 2001. Update 2006.

Sebastian L. Johnston
Department of Respiratory Medicine
National Heart and Lung Institute
Wright Fleming Institute of Infection & Immunity
Imperial College London, Norfolk Place
London W2 1PG (UK)
Tel. +44 20 7594 3764, Fax +44 20 7262 8913, E-Mail s.johnston@imperial.ac.uk

Sjöbring U, Taylor JD (eds): Models of Exacerbations in Asthma and COPD.
Contrib Microbiol. Basel, Karger, 2007, vol 14, pp 113–125

Animal Models of Cigarette Smoke-Induced Chronic Obstructive Lung Disease

Andrew Churg, Joanne L. Wright

Department of Pathology, University of British Columbia, Vancouver, B.C., Canada

Abstract

Recent years have seen an explosion of animal models of cigarette smoke-induced chronic obstructive lung disease (COPD). Almost all of these have concentrated on the induction and prevention of emphysema. Neutrophils and neutrophil elastase, macrophages and macrophage-derived metalloproteases, lymphocytes, TNF-α, and oxidants have all been shown to play a role in the pathogenesis of emphysema in animal models, and interventions using either knockout mice or drugs have indicated possible preventive/therapeutic avenues. There is less in the way of models of smoke-induced small airway remodeling and almost nothing is known of its pathogenesis. Cigarette smoke has been shown to induce vascular remodeling and pulmonary hypertension in laboratory animals, and these mechanisms are beginning to be understood. A major limitation of existing animal models is that most produce relatively mild disease (no more severe than corresponding to the GOLD 2 stage of human COPD), and none of the models show the smoke-independent progressive disease seen in humans with GOLD 3 or 4 COPD. There are no models of cigarette smoke-induced chronic bronchitis in animals and there are no models of acute exacerbations of COPD.

Introduction

Chronic obstructive pulmonary disease (COPD) is now the fourth leading cause of death in Europe and North America, and produces extensive morbidity as well, with considerable cost to health care systems worldwide. Despite extensive analysis of lungs from humans with COPD, there was until recently a surprising dearth of animal models of the various components of this disease. We produced what in retrospect was the first laboratory animal model when we showed in 1990 that guinea pigs exposed to cigarette smoke on a daily basis developed both emphysema and pulmonary hypertension [1]; in the last 10 years an increasing number of investigations using chronic cigarette smoke exposure of rats, mice, and guinea pigs have been published.

This paper briefly reviews the findings of these chronic exposure models. To correspond to human disease, we have divided the models into those that produce emphysema, those that produce small airway remodeling (SAR), and those that produce pulmonary hypertension. We limited this review to models in which anatomic changes (mostly emphysema) are found after smoke exposure, which in practice means relatively long-term exposures.

It is important to note at the outset a major limitation of almost all existing animal models; namely, that the disease they produce is generally relatively mild, equivalent at most to the morphological changes seen in human GOLD (Global Initiative on Chronic Obstructive Lung Disease [2]) stage 1 or 2 disease. This is a serious limitation, because the majority of morbidity and mortality in COPD occurs in patients with GOLD stage 3 or 4 disease (i.e., FEV_1 <50% predicted), but the type of severe emphysema or severe SAR found in GOLD stage 3 or 4 patients simply cannot be reproduced in animals. In part this problem may arise from the practicalities of producing animal models: a 6-month experiment, which is generally required to observe any significant emphysema in mice or guinea pigs, is long and expensive in terms of research budgets, but even extending the smoking period to 12 months does not produce dramatic disease in guinea pigs [1]. Moreover, in advanced GOLD stage patients, COPD typically progresses even after the patient stops smoking, but this phenomenon has not been reproduced in animal models.

Models of Emphysema: Pathogenesis and Interventions

Anatomic Patterns of Emphysema
Most animal models of emphysema produce dilated alveolar ducts (fig. 1), an anatomic pattern reasonably similar to the centrilobular emphysema that is the most common form found in human cigarette smokers. Studies from our laboratory using scanning electron microscopy suggest that in fact the parenchyma between the alveolar ducts is also abnormal, with increases in size and number of the pores of Kohn that connect adjacent alveoli [3]; this phenomenon is again similar to what has been found in the lungs of human cigarette smokers [4, 5]. We have also demonstrated that while there is an initial decrease in alveolar collagen shortly after the commencement of smoke exposure, long-term exposure increases the amount of alveolar collagen fibers [6].

Since the lesions produced in laboratory animals are never large enough to see with the naked eye, and can be subtle even on microscopic examination, the

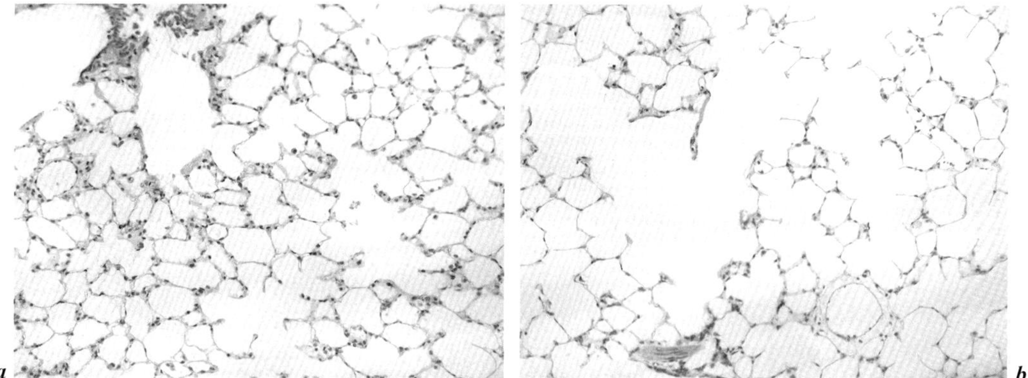

Fig. 1. Comparison of lung parenchyma between control guinea pigs (*a*) and those exposed to cigarette smoke (*b*) for 6 months. Note the enlarged alveolar ducts and alveoli in the smoke-exposed animals.

severity of emphysema is assessed by a morphometric measure of airspace size, most commonly mean linear intercept (Lm), or surface to volume ratio [see 7 for a review]. The increase that is found in Lm varies dramatically from laboratory to laboratory, and appears to be dependent on the type of cigarette used (the older high tar cigarettes such as University of Kentucky 2R1s produce more severe lesions than newer low tar or filtered cigarettes), smoking machine (each machine delivers quite different amounts of smoke), smoking regime, and species; thus each laboratory must do careful controls, and comparisons among laboratories are not straightforward; indeed, different laboratories smoking the same strain of mice for the same length of time report not only differences in Lm but differences in the amount of inflammation [compare for example 8 and 9].

In our experience guinea pigs are generally more sensitive to the effects of smoke than are mice, although this varies considerably by mouse strain (see below), and rats are the least susceptible. We generally find an increase in Lm of about 30–40% after 6 months of smoke exposure. Arguably one might do better to measure only alveolar ducts, since that is the major site of the abnormal lesion, and when we looked at the volume proportion of alveolar duct air in C57Bl/6 mice exposed to smoke for 6 months, we found an 88% increase, compared to a 38% increase for Lm [8]. However, measurements of alveolar duct air present greater morphometric problems.

While most reports of mice exposed to smoke have described predominantly alveolar duct enlargement, Takubo et al. [10] found that mice genetically deficient in α_1-antitrypsin (pallid mice) developed a pattern of airspace enlargement that was reasonably similar to human panlobular emphysema; this anatomic abnormality was accompanied by measurable increases in compliance, as opposed to C57Bl/6 mice which had enlarged alveolar ducts but no compliance changes. Emphysema in the pallid mice developed earlier than in the C57 mice, but was no more severe at 6 months than in the C57 mice (in terms of increases in Lm). The pallid mice but not the C57 mice had a T cell inflammatory infiltrate in the alveolar walls. Cavarra et al. [11] also found that pallid mice developed emphysema in a panlobular pattern. Interpretation of these findings is not straightforward because pallid mice develop a pattern of mild panlobular emphysema even without smoke exposure [10, 11], but, since α_1-antitrypsin is the major anti-(serine) protease in the lower respiratory tract, in many senses they do support the original protease-antiprotease hypothesis of smoke-induced emphysema (see below).

Interventions

Serine Protease Inhibitors
The observations that humans deficient in α_1-antitrypsin developed early onset emphysema, particularly if they smoked, and that emphysema could be produced in laboratory animals by instillation of elastases, lead to the protease-antiprotease hypothesis. In short this says that smoke evokes an inflammatory infiltrate in the lower respiratory tract, and that these inflammatory cells release proteases that overwhelm the local antiproteolytic defenses, allowing proteolytic degradation of the alveolar wall matrix and the eventual development of emphysema.

The protease-antiprotease hypothesis is generally accepted (although it has been challenged by other theories such as failure of lung maintenance [12]), but the crucial inflammatory cells/proteases are the subject of considerable debate. In its original formulation, the neutrophil and neutrophil-derived proteases, particularly neutrophil elastase, were considered the major culprits. Although some investigators now downplay the neutrophil elastase in favor of macrophage-derived metalloproteases, animal models of emphysema show clearly that the neutrophil is involved. The fact that animals deficient in α_1-antitrypsin develop more severe emphysema after smoke exposure compared to animals with normal α_1-antitrypsin levels has been alluded to above. Conversely, exogenous augmentation of lower respiratory tract anti-serine protease levels provides partial protection (up to 70%) against smoke-induced emphysema (table 1): this has been shown for both α_1-antitrypsin itself, whether administered by intraperitoneal injection [13] or inhalation [14], and a synthetic neutrophil elastase inhibitor [15]. Further, mice with targeted deletion of the neutrophil elastase gene also are partially (59%) protected against emphysema [16] (table 1). Sixty to 70% protection against emphysema should not be regarded as a useless

Table 1. Protection against emphysema in various mouse models (measured as protection against increases in Lm)

	Emphysema protection, %
Serine elastase inhibitors/interference with neutrophil elastase	
Neutrophil elastase knockout (mice) [16]	59
α_1-Antitrypsin (injected; mice) [13]	63
α_1-Antitrypsin (inhaled; mice) [14]	72
ZD0892 (guinea pigs) [15]	45
Metalloprotease inhibitors/interference with metalloproteases	
MMP inhibitor GM6001 (mice) [35]	90
MMP-12 knockout [15, 17]	100
MMP inhibitor RS113456 (mice) [20][a]	100
MMP inhibitor RS132908 (mice) [20]	75
MMP inhibitor CP-471,474 at 2 months[b] (guinea pigs) [19]	100
MMP inhibitor CP-471,474 at 4 months[b] (guinea pigs) [19]	30
Selective MMP-9/MMP-12 inhibitor (Churg unpubl.; guinea pigs)	70
Anti-inflammatory treatments/genetic manipulations	
TNF-α receptor I/II knockout (mice) [8]	71
PDE4 inhibitor roflumilast (mice) [25]	100

[a]Started after 3 months of smoke exposure.
[b]Alveolar area measured.

result, since that degree of protection in humans would probably be of considerable clinical benefit, but the fact that complete protection is not present indicates that other mechanisms beyond just serine proteases are involved.

Metalloproteases

In the last 15 years there has been increasing recognition that matrix metalloproteases can degrade elastin, and increasing documentation of the fact that human lungs with emphysema and alveolar macrophages derived from these lungs produce enhanced amounts of a variety of metalloproteases [reviewed in 17]. The most convincing evidence for a role for metalloprotease in emphysema comes from the work of Shapiro and colleagues [18] who showed that mice with targeted deletion of MMP-12 (macrophage metalloelastase) were 100% protected against cigarette smoke-induced emphysema (table 2). Studies using synthetic metalloprotease inhibitors have also provided protection, but of more variable degree (table 2).

Exactly which metalloproteases are important in humans is an open question, particularly since the role(s) of cognate metalloproteases varies considerably across species, and not all species have the same metalloproteases; mice and rats,

Table 2. Effects of antioxidant interventions on Lm after chronic smoke exposure

	Increase or protection
Transgenic CuZnSOD mice [22]	100% protection
Nrf2 knockout mice [23]	314% greater than wild type (6 months)
Nrf2 knockout mice [24]	53% greater than wild type (6 months)
High antioxidant levels (ICR mice) [11]	9% increase (not statistically significant) at 7 months

for example, lack a true MMP-1, a collagenase. Selman et al. [19] found only minimal protection (30%) at 4 months using a broad spectrum metalloprotease inhibitor in guinea pigs; however, we observed that a selective MMP-9/MMP-12 inhibitor provided about 70% protection in guinea pigs after 6 months of smoke exposure [Churg et al., unpubl. data] (table 1). This finding is of particular interest, since it suggests that MMP-9 and MMP-12 are crucial to the development of emphysema in a nonmurine species, implying that they may also be of major importance in humans. An additional important observation was made by Martin et al. [20] who showed that late intervention with a broad spectrum metalloprotease inhibitor, RS113456, prevented further progression of emphysema in mice, but we were unable to show any benefit of late intervention with the serine elastase inhibitor, ZD0892. Since humans are likely to be treated at a relatively late stage (or at least not from their first cigarette), a treatment that arrests emphysema progression will be crucial.

Oxidant Protection or Enhanced Sensitivity

Cigarette smoke is a highly concentrated source of oxidants, both reactive oxygen species and reactive nitrogen species. There is considerable evidence for oxidative stress in cigarette smokers [21], but the exact contribution of oxidants and oxidative damage to emphysema in humans is unclear. However, recent animal data suggest that oxidant protection does play an important role (table 2). Cavarra et al. [11] showed that ICR mice, which increased their levels of lavage antioxidant protection, did not develop emphysema, whereas C57Bl/6J and DBA/2, which decreased their levels of antioxidant protection, did. Foronjy et al. [22] reported that mice transgenic for human CuZnSOD were completely protected against emphysema and showed decreased neutrophil influx as well as decreased levels of the oxidation markers 3-nitrotyrosine and 8-hydrodeoxyguanosine. Ragasamy et al. [23] and Iizuka et al. [24] utilized mice with targeted deletion of Nrf2, a transcription factor that induces a variety of detoxifying and antioxidant elements under the control of the antioxidant response element. Both sets of authors found that Nrf2–/– mice developed earlier and more severe emphysema compared to wild-type controls, along with increased evidence of oxidative stress and increased lavage neutrophil activity. A particularly interesting observation was the fact that both α_1-antitryspin and secretory leukoprotease inhibitor production were decreased in Nrf2 mice, raising the question of whether the crucial process is oxidant protection or serine protease protection.

Inflammatory Mediators

As noted above, most investigators believe that persisting inflammation is the crucial driving force behind emphysema and where this has been measured, the antiproteolytic and antioxidant interventions noted above have generally reduced smoke-evoked inflammation.

A number of papers have been published that specifically address inflammatory cell chemoattractants in long-term models. Churg et al. [8] showed that mice lacking TNF-α receptors I and II were about 70% protected against emphysema compared to wild type after 6 months of smoke, implying that TNF-α-driven inflammation is important in emphysema. In addition, the knockout mice were completely protected against smoke-induced increases in MMP-2 and MMP-9, but only partially protected against MMP-12, 13 or 14, suggesting that these enzymes might be responsible for the remaining 30% of airspace enlargement.

Phosphodiesterases (PDE) are enzymes that degrade cyclic nucleotides. PDE4 in particular degrades cyclic $3',5'$-adenosine monophosphate, an anti-inflammatory substance. Martorana et al. [25] treated mice with the PDE4 inhibitor, roflumilast, and showed that the drug completely protected against emphysema in a dose-response fashion, and decreased, but did not completely abrogate, macrophage influx.

Statins are 3-hydroxy-3-methyl-glutaryl coenzyme A (HMG-CoA) reductase inhibitors that are generally used as lipid-lowering agents. Statins also have a number of other properties including anti-inflammatory and antioxidant effects. Lee et al. [26] reported that treatment with simvastatin completely abrogated cigarette smoke-induced emphysema in Sprague-Dawley rats and also reduced peribronchial and perivascular inflammation. However, the authors claimed a 74% increase in Lm after 4 months of smoking; in our experience it is impossible to achieve this increase in Lm in rats (or for that matter in any other genetically intact species), even with 6 or 12 months of smoke exposure, and we suggest that enthusiasm for using this drug as an intervention be tempered until its effects have been firmly established.

Other Strain Differences in Mice

It is apparent from the comments above that strain differences in mice play a major role in the development or failure of development of emphysema

Table 3. Increases in Lm by mouse strain compared to nonexposed mice of the same strain

	Increase in Lm, %
ICR [11]	9 (not statistically significant; 7 months)
DBA/2 [11]	26 (7 months)
C57Bl/6 [11]	14 (7 months)
NZW [9]	−11 (not statistically significant; 6 months)
C57Bl/6 [9]	13.2 (6 months)
A/J [9]	17.9 (6 months)
SJ/L [9]	23.2 (6 months)
AKR/J [9]	38.0 (6 months)

(table 3). Guerassimov et al. [9] exposed 5 different strains, NZWLac/J, C57BL6/J, A/J, SJ/L, and AKR/J, to smoke for 6 months and found that they could be divided into three groups: resistant strains (NZW) that showed no increase in either Lm or compliance, partially susceptible strains (C57BL6/J, A/J, and SJ/L) that showed an increase in Lm but not in compliance, and a highly susceptible strain (AKR/J) that demonstrated increases in Lm, compliance, inflammatory cells, and Th1 cytokines. These findings support the idea of a genetic basis for cigarette smoke susceptibility, as do the data of Cavarra et al. [11] mentioned above. The results of Guerassimov et al. may also imply that coinfections are important mediators of COPD development, since ARK/J mice have an integrated viral genome and are naturally viremic (and in fact all develop hematopoietic malignancies).

Autoimmunity or Acquired Immunity to Tobacco Constituents as a Cause of Emphysema

A recent hypothesis proposes that progression in emphysema may be mediated by the development of autoimmunity to smoke-modified (oxidized) proteins or to matrix fragments resulting from smoke-induced inflammation, and/or the development of antibodies against one or more smoke constituents. Van der Strate et al. [27] reported that B cell follicles with an oligoclonal, antigen-specific reaction were present in C57 mice chronically exposed to cigarette smoke, and that the development of the follicles was progressive over time and correlated with the amount of emphysema. However, there is no evidence that the disease of these animals progresses after stopping smoke exposure, again emphasizing the failure of laboratory animal models to recapitulate severe COPD in humans.

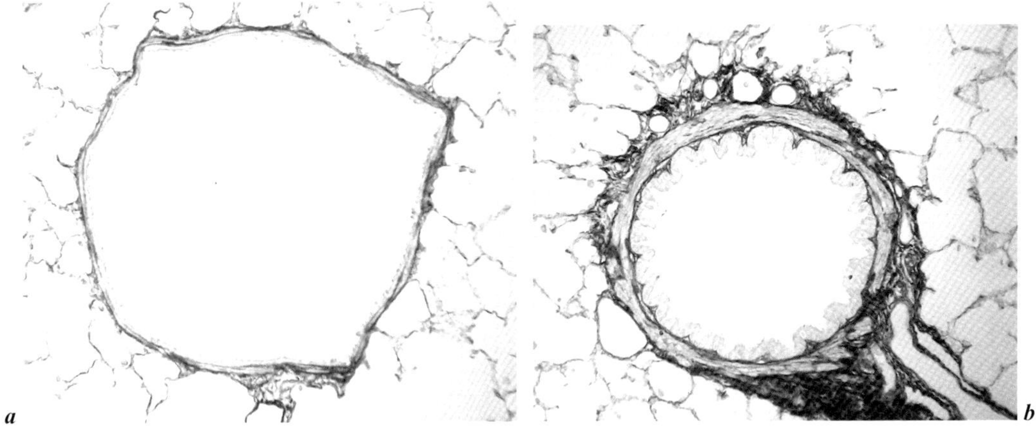

Fig. 2. Comparison of airways between control guinea pigs (*a*) and those exposed to cigarette smoke (*b*) for 6 months using a picrosirius red stain. Collagen appears black. Note the increase in collagen and smooth muscle in the smoke-exposed animals.

Models of SAR (Small Airway Disease)

SAR is an important cause of airflow obstruction in humans with COPD, but little is known about SAR in animal models of chronic smoke exposure. Whereas it is possible to see emphysema on casual observation of microscopic sections from these animals, in our experience one cannot reliably separate the small airways from smoke-exposed and control animals by simple examination, although as in humans, individual airways can demonstrate significant structural remodeling (fig. 2). However, morphometric measurements of the airway wall indicate that SAR does indeed occur in the mouse and guinea pig, and further that the major abnormality in guinea pigs is an increase in collagen rather than muscle [Wright, unpubl. observations]. Of note, the selective dual MMP-9/MMP-12 inhibitor mentioned above completely prevented the development of SAR in guinea pigs [Churg, unpubl. observations].

Models of Pulmonary Hypertension

Remodeling of the pulmonary arterial vasculature occurs in human smokers, and is characterized by intimal fibrosis, muscular hypertrophy and adventitial fibrosis in the usually muscularized arteries, and by extension of smooth

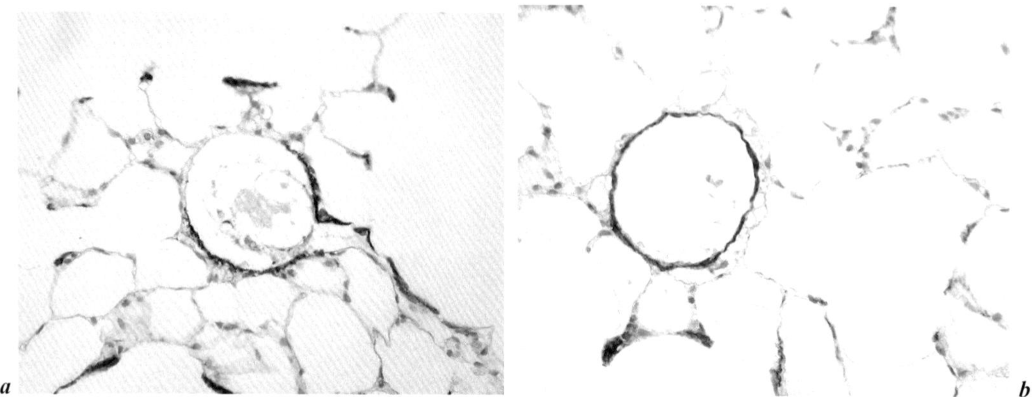

Fig. 3. Comparison of the small, normally poorly muscularized vessels adjacent to the alveolar ducts between control guinea pigs (*a*) and those exposed to cigarette smoke for 6 months (*b*) using a smooth muscle actin immunohistochemical stain. Note that the vessels in the smoke-exposed animals are fully muscularized.

muscle into the usually partially muscularized vessels causing them to become completely muscularized.

The guinea pig animal model demonstrates vascular remodeling with chronic smoke exposure, and this is associated with increased pulmonary arterial pressure [28] (fig. 3). We have shown that the increased pulmonary arterial pressure cannot be ascribed to capillary destruction [29], but instead appears to be related to direct smoke-mediated increases in production of vasoactive mediators, including endothelin and VEGF, in the vessel walls [30] and subsequent vessel muscularization. We have recently demonstrated increased metalloproteinase mRNA, protein, and activity in the vessels of mice exposed short-term to cigarette smoke [31]. Interestingly, administration of the serine elastase inhibitor ZD0892 reduced vascular remodeling [32] in the guinea pig.

Models of Chronic Bronchitis and Exacerbations

Compared to humans, bronchial glands are sparse in mice, rats, and guinea pigs, and they are generally limited to the proximal portion of the trachea. As well, there is much less of a tendency for these animals to develop goblet cell metaplasia with smoke exposure, although this is somewhat strain and species dependent. Guinea pigs do exhibit goblet cell metaplasia of at least a modest degree [33], whereas C57 mice have only minimal alterations [25], and in our experience, rats have only slightly greater metaplasia in the small airways than C57 mice.

Interestingly, in the guinea pig, smoking cessation reduces the metaplasia [34]. The relative paucity of bronchial glands and goblet cells is probably responsible for the failure of any investigators to produce a model of human chronic bronchitis.

Along this line, the other major lack is the absence of models of acute exacerbation. This is an important phenomenon in humans and a major source of both morbidity and mortality. The problem is most likely related to the mild disease that can be produced in animals, because acute exacerbations are typically seen in humans with advanced COPD, and may also be related to the presence of excessive mucus production.

Conclusions

During recent years a number of in vivo models to study the pulmonary changes evoked by chronic exposure to cigarette smoke, the predominant cause of COPD, have been developed. From these models a role in emphysema development for different immune cell types, such as neutrophils, macrophages and lymphocytes, as well as for various mediators, including neutrophil elastase, macrophage-derived metalloproteases, and TNF-α, have been established. In contrast, up to now these models have supplied limited information with respect to SAR, a distinguishing feature of COPD in humans. The challenge for further development of in vivo smoke models will be to incorporate features outside emphysema development, including SAR, excessive mucus production and exacerbation.

Acknowledgment

This study was supported by grants MOP 42539 and 81409 from the Canadian Institutes of Health Research.

References

1   Wright JL, Churg A: Cigarette smoke causes physiological and morphological changes of emphysema in the guinea pig. Am Rev Respir Dis 1990;142:1422–1428.
2   Pauwels RA, Buist AS, Ma P, Jenkins CR, Hurd SS; GOLD Scientific Committee: Global strategy for the diagnosis, management, and prevention of chronic obstructive pulmonary disease: National Heart, Lung, and Blood Institute and World Health Organization Global Initiative for Chronic Obstructive Lung Disease (GOLD): executive summary. Respir Care 2001;46:798–825.
3   Wright JL: The importance of ultramicroscopic emphysema in cigarette smoke induced lung disease. Lung 2001;179:71–81.
4   Cosio MG, Shiner RJ, Saetta M, Wang N-S, King M, Ghezzo H, Angus E: Alveolar fenestrae in smokers. Am Rev Respir Dis 1986;133:126–131.

5   Kuhn C, Tavassoli F: The scanning electron microscopy of elastase-induced emphysema. Lab Invest 1976;34:2–9.
6   Wright JL, Churg A: Smoke-induced emphysema in the guinea pig is associated with morphometric evidence of collagen breakdown and repair. Am J Physiol 1995;268:L17–L20.
7   Thurlbeck WM: Christie Lecture: emphysema then and now. Can Respir J 1994;1:21–39.
8   Churg A, Wang RD, Tai H, Wang X, Xie C, Wright JL: Tumor necrosis factor-α drives 70% of cigarette smoke-induced emphysema in the mouse. Am J Respir Crit Care Med 2004;170:492–498.
9   Guerassimov A, Hoshino Y, Takubo Y, Turcotte A, Yamamoto M, Ghezzo H, Triantafillopoulos A, Whittaker K, Hoidal JR, Cosio MG: The development of emphysema in cigarette smoke-exposed mice is strain dependent. Am J Respir Crit Care Med 2004;170:974–980.
10 Takubo Y, Guerassimov A, Ghezzo H, Triantafillopoulos A, Bates JH, Hoidal JR, Cosio MG: Alpha1-antitrypsin determines the pattern of emphysema and function in tobacco smoke-exposed mice: parallels with human disease. Am J Respir Crit Care Med 2002;166:1596–1603.
11 Cavarra E, Bartalesi B, Lucattelli M, Fineschi S, Lunghi B, Gambelli F, Ortiz LA, Martorana PA, Lungarella G: Effects of cigarette smoke in mice with different levels of alpha(1)-proteinase inhibitor and sensitivity to oxidants. Am J Respir Crit Care Med 2001;164:886–890.
12 Tuder RM, McGrath S, Neptune E: The pathobiological mechanisms of emphysema models: what do they have in common? Pulm Pharmacol Ther 2003;16:67–78.
13 Churg A, Wang RD, Xie C, Wright JL: α-1-Antitrypsin ameliorates cigarette smoke-induced emphysema in the mouse. Am J Respir Crit Care Med 2003;168:199–207.
14 Pemberton PA, Kobayashi D, Wilk BJ, Henstrand JM, Shapiro SD, Barr PJ: Inhaled recombinant alpha-1-antitrypsin ameliorates cigarette smoke-induced emphysema in the mouse. COPD 2006;3: 101–108.
15 Wright JL, Farmer S, Churg A: A synthetic serine elastase inhibitor reduces cigarette smoke induced emphysema in guinea pigs. Am J Respir Crit Care Med 2002;166:954–960.
16 Shapiro SD, Goldstein NM, Houghton AM, Kobayashi DK, Kelley D, Belaaouaj A: Neutrophil elastase contributes to cigarette smoke-induced emphysema in mice. Am J Pathol 2003;163: 2329–2335.
17 Churg A, Wright J: Proteases and emphysema. Curr Opin Pulm Med 2005;11:153–159.
18 Hautamaki R, Kobayashi D, Senior R, Shapiro S: Requirement for macrophage elastase for cigarette smoke-induced emphysema in mice. Science 1997;277:2002–2004.
19 Selman M, Cisneros-Lira J, Gaxiola M, Ramirez R, Kudlacz EM, Mitchell PG, Pardo A: Matrix metalloproteinases inhibition attenuates tobacco smoke-induced emphysema in guinea pigs. Chest 2003;123:1633–1641.
20 Martin RL, Shapiro SD, Tong SE, Van Wart H: Macrophage metalloelastase inhibitors; in Hansel TT, Barnes PJ (eds): New Drugs for Asthma, Allergy and COPD. Prog Respir Res. Basel, Karger, 2001, vol 31, pp 177–180.
21 MacNee W: Pulmonary and systemic oxidant/antioxidant imbalance in chronic obstructive pulmonary disease. Proc Am Thorac Soc 2005;2:50–60.
22 Foronjy RF, Mirochnitchenko O, Propokenko O, Lemaitre V, Jia Y, Inouye M, Okada Y, D'Armiento JM: Superoxide dismutase expression attenuates cigarette smoke- or elastase-generated emphysema in mice. Am J Respir Crit Care Med 2006;173:623–631.
23 Rangasamy T, Cho CY, Thimmulappa RK, Zhen L, Srisuma SS, Kensler TW, Yamamoto M, Petrache I, Tuder RM, Biswal S: Genetic ablation of Nrf2 enhances susceptibility to cigarette smoke-induced emphysema in mice. J Clin Invest 2004;114:1248–1259.
24 Iizuka T, Ishii Y, Itoh K, Kiwamoto T, Kimura T, Matsuno Y, Morishima Y, Hegab AE, Homma S, Nomura A, Sakamoto T, Shimura M, Yoshida A, Yamamoto M, Sekizawa K: Nrf2-deficient mice are highly susceptible to cigarette smoke-induced emphysema. Genes Cells 2005;10: 1113–1125.
25 Martorana PA, Beume R, Lucattelli M, Wollin L, Lungarella G: Roflumilast fully prevents emphysema in mice chronically exposed to cigarette smoke. Am J Respir Crit Care Med 2005;172: 848–853.
26 Lee J-H, Lee D-S, Kim E-K, Choe K-H, Oh Y-M, Shim T-S, Kim S-E, Lee Y-S, Lee S-D: Simvastatin inhibits cigarette smoking-induced emphysema and pulmonary hypertension in rat lungs. Am J Respir Crit Care Med 2005;172:987–993.

27 van der Strate BW, Postma DS, Brandsma CA, Melgert BN, Luinge MA, Geerlings M, Hylkema MN, van den Berg A, Timens W, Kerstjens HA: Cigarette smoke-induced emphysema: a role for the B cell? Am J Respir Crit Care Med 2006;173:751–758.

28 Wright JL, Churg A: Effect of long-term cigarette smoke exposure on pulmonary vascular structure and function in the guinea pig. Exp Lung Res 1991;17:997–1009.

29 Yamato H, Sun J-P, Churg A, Wright JL: Guinea pig pulmonary hypertension caused by cigarette smoke cannot be explained by capillary bed destruction. J Appl Physiol 1997;82:1644–1653.

30 Wright JL, Tai H, Churg A: Vasoactive mediators and pulmonary hypertension after cigarette smoke exposure in the guinea pig. J Appl Physiol 2006;100:672–678.

31 Wright JL, Tai H, Wang R, Wang X, Churg A: Cigarette smoke upregulates pulmonary vascular matrix metalloproteinases via TNFα signaling. Am J Physiol 2007;292:L125–L133. DOI: 10.1152/ ajplung.00539.2005.

32 Wright JL, Farmer S, Churg A: A neutrophil elastase inhibitor reduces cigarette smoke-induced remodeling of lung vessels. Eur Respir J 2003;22:77–81.

33 Wright JL, Ngai T, Churg A: Effect of long-term exposure to cigarette smoke on the small airways of the guinea pig. Exp Lung Res 1992;18:105–114.

34 Wright JL, Churg A: Smoking cessation decreases the number of metaplastic secretory cells in the small airways of the guinea pig. Inhal Toxicol 2002;14:101–107.

35 Pemberton P, Cantwell JS, Kim K, Sundin DJ, Kobayashi D, Fink JB, Shapiro SD, Barr PJ: An inhaled matrix metalloprotease inhibitor prevents cigarette smoke-induced emphysema in the mouse. COPD 2005;3:303–310.

Andrew Churg, MD
Department of Pathology, University of British Columbia
2211 Wesbrook Mall
Vancouver, BC V6T 2B5 (Canada)
Tel. +1 604 822 7775, Fax +1 604 822 7635, E-Mail achurg@interchange.ubc.ca

Sjöbring U, Taylor JD (eds): Models of Exacerbations in Asthma and COPD.
Contrib Microbiol. Basel, Karger, 2007, vol 14, pp 126–141

Animal Models of Chronic Obstructive Pulmonary Disease Exacerbations

Gordon J. Gaschler[a], *Carla M.T. Bauer*[a], *Caleb C.J. Zavitz*[a],
Martin R. Stämpfli[a,b]

[a]Department of Pathology and Molecular Medicine, Centre for Gene Therapeutics and
[b]Department of Medicine, McMaster University, Hamilton, Ont., Canada

Abstract

Modeling acute exacerbations of chronic obstructive pulmonary disease (AECOPD) in animals has proven challenging due to the clinical and pathological complexity of the underlying disease. This has hindered the progress in understanding the cellular and molecular mechanisms that lie beneath AECOPD. In this chapter, we will address modeling possibilities of AECOPD that may be drawn from the current knowledge of factors that cause exacerbations. Importantly, since it is widely accepted that the most common causes of AECOPD are viral and bacterial infections, animal models of AECOPD should incorporate both the causative agents of exacerbation: viruses and bacteria. However, other factors that are also believed to determine both progression of COPD, as well as the frequency and severity of AECOPD, such as proteolytic enzymes, cigarette smoke or other noxious stimuli must also be considered. Such animal models will provide mechanistic insight into the etiology of AECOPD and will prove invaluable in furthering our understanding of key events in disease pathogenesis.

Introduction

Animal models are important tools for elucidating mechanisms of disease pathogenesis, for identifying novel therapeutic targets, and for preclinical screening of intervention strategies. Modeling acute exacerbations of chronic obstructive pulmonary disease (AECOPD) in animals has proven challenging due to the clinical and pathological complexity of the underlying disease. This has hindered the progress in understanding the cellular and molecular mechanisms underlying AECOPD. In this chapter, we will address modeling possibilities that may be drawn from the current knowledge of factors that cause exacerbations, and determinants of exacerbation frequency and severity.

COPD Exacerbation

Clinically, AECOPD is viewed as a sustained worsening of symptoms from the patient's stable COPD state, beyond normal day-to-day variations, which necessitate a change in medication [1]. Symptoms may include increased dyspnea, wheezing, chest tightness, cough, and changes in the level, color and/or tenacity of sputum [2]. Unfortunately, this clinical definition of AECOPD is of limited guidance for the development of animal models.

COPD is not a single disease but rather a syndrome whose functional and clinical hallmark is progressive and largely irreversible airflow limitation [2]. In an individual patient, varying degrees of chronic bronchitis, bronchiolitis, and emphysema contribute to this chronic airflow limitation [3–5]. Various approaches have been used to model the histopathological features of COPD. These have included exposing animals to cigarette smoke or inflammatory stimuli, instilling proteolytic enzymes into the airways, and the study of spontaneous gene mutants and knockout strains [reviewed in 6–10]. Typically, these studies have addressed single aspects of COPD, such as inflammation, emphysema, or mucus production, but have failed to recapitulate the entirety of the COPD syndrome. Because we lack a comprehensive model of COPD, it is impossible to model AECOPD as a worsening of COPD. As a solution to this inherent problem, we have to consider the individual factors that are associated with COPD and AECOPD, and subsequently incorporate them either independently or in combination into animal models.

It is widely accepted that the most common causes of AECOPD are viral and bacterial infections [11–14]. Hence, animal models of AECOPD should incorporate the causative agents of exacerbation, namely viruses and bacteria. In addition other factors that may both be determinants for the progression of stable disease, and in some cases probably also of AECOPD, such as noxious stimuli, including cigarette smoke and susceptibility factors, such as genetic perturbation of the protease-antiprotease balance, need also to be included.

Determinants of Exacerbation Frequency and Severity

COPD is found almost exclusively in smokers; over 80% of all cases are attributable to active smoking, and an additional 10% are believed to be due to exposure to environmental tobacco smoke [15]. Notably, patients with COPD, especially in the early stages of disease, typically continue to smoke. These patients often suffer from greater sputum production and increased episodes of coughing than do patients who cease smoking [16], which is understood to be associated with more frequent exacerbations. In view of this, cigarette smoking is an important component of AECOPD models, not only as an etiologic factor for COPD itself, but also as a determinant of exacerbation frequency and severity. Addressing the effect of cigarette smoke exposure on exacerbation-causing bacterial and viral infections in animals may provide insight into the mechanisms underlying AECOPD and may lead to the development of more complete models.

In addition to the association between smoking and increased risk of viral and bacterial infection, clinical evidence shows a relationship between lower airway bacterial colonization and exacerbation frequency [17]. Why microbial agents colonize the lower respiratory tract, normally a sterile compartment, is unclear and will be discussed in more detail later in this chapter. Similarly, bacterial colonization may impact the susceptibility to secondary viral and bacterial infections, or alter the ensuing immune-inflammatory responses, through mechanisms that are still poorly understood.

The severity of the stable COPD state is an important factor of exacerbation frequency. It has been postulated that changes within the diseased lung such as increased mucus production, thickening of the epithelium, or changes in epithelial cell integrity, and parenchymal damage, all of which become more marked with disease progression, may predispose COPD patients to microbial infection [18]. Similarly, the lung function in patients suffering from COPD may have deteriorated to an extent that relatively minor changes in lung function brought on by microbial infections may be clinically manifested as exacerbations. Complicating these issues is the fact that relatively little is known about how viral and bacterial infections affect lung function. Thus, understanding the underlying pathology of viral and bacterial infections and their consequences to lung function both in healthy and diseased lungs becomes increasingly important to the study of mechanisms leading to AECOPD.

Modeling Bacterial Exacerbations

Bacterial infections have been estimated to cause half of all AECOPD [12]. The most commonly isolated bacteria during episodes of AECOPD are nontypeable *Haemophilus influenzae*, *Moraxella catarrhalis* and *Streptococcus pneumoniae*, while in patients with more severe disease progression, invasive bacteria such as *Pseudomonas aeruginosa* or *Chlamydia pneumoniae* may be isolated [12]. As previously mentioned, even during stable disease bacteria are frequently detected in the lungs of COPD patients [19–21]. This has made the precise role of bacterial infections in AECOPD controversial, as bacteria isolated during episodes of AECOPD may either be from infection or colonization. However, recent studies have elegantly shown that AECOPD in colonized individuals is often associated with the acquisition of a new strain of bacteria [22], which is also linked to the appearance of antibodies specifically directed against the new bacterial strain [23]. Animal models of bacterial infection or colonization in AECOPD should therefore include not only drivers of COPD such as cigarette smoke exposure and persistent bacterial colonization, but also strains of bacteria that are capable of causing exacerbation.

Cigarette Smoke and Bacterial Agents

Cigarette smoke has been demonstrated to compromise various aspects of the immune system. For example, cigarette smoke can cause impaired mucociliary clearance [24], decreased levels of both the phospholipid fraction and surfactant-associated proteins A and D of the surfactant system [25, 26], impaired alveolar macrophage phagocytosis, killing and cytokine induction [27–36], and disruption of the epithelial cell barrier [37, 38]. All of these consequences of exposure to smoke may contribute to an impaired antibacterial immune response in smokers, and by extension, to bacterial exacerbation of COPD.

Numerous studies have examined the effect of cigarette smoke exposure on responses to bacterial antigens. More specifically, mediators of the adaptive immune response such as the ex vivo production of Th1-type cytokines have been shown to be attenuated by smoke [34, 35]. In contrast the inflammatory response to an innate stimuli such as lipopolysaccharide may be heightened in vivo [39]. These studies serve to identify some of the pathways affected by cigarette smoke, elucidate mechanisms contributing to chronic inflammation, and provide a basic scientific framework to develop and test therapeutic targets. The limitation of these experimental approaches, however, is that they ignore the intricacy associated with infection by live, replication-competent bacteria. During bacterial infections numerous bacterial antigens are present, including constituents of bacterial cell walls such as peptidoglycan and lipopolysaccharide, as well as lipoproteins, lipoteichoic acids, flagella, and toxins. These antigens stimulate pattern recognition receptors of resident innate cells, including Toll-like and Nod-like receptors, resulting in the production of proinflammatory cytokines and chemokines. In turn, neutrophils and monocytes/ macrophages are recruited from the circulation and activated, resulting in local inflammation. Accordingly, the immunosuppressive effects of cigarette smoke may in fact propagate inflammatory processes by delaying bacterial clearance, driving tissue pathology and progressive airway obstruction.

As early as the 1960s, in vivo studies demonstrated that bacterial clearance was delayed from the airways of cigarette smoke-exposed animals [40–43]. This is indicative of cigarette smoke predisposing to bacterial infection or colonization. Work previously done in our lab with *P. aeruginosa* has shown that mice exposed to cigarette smoke demonstrate a delayed rate of bacterial clearance as compared to sham-exposed mice [44]. This study also demonstrated that delayed clearance was associated with a skewing of the innate inflammatory response: the increased bacterial burden was associated with *increased* airway and tissue inflammation *despite* evidence of suppressed alveolar macrophage function. Furthermore, we observed increased levels of proinflammatory cytokines and chemokines, myeloperoxidase, proteases, and perhaps most importantly, these changes were associated with a significant deterioration of the animals' health status. In this model, exposure to cigarette smoke alone did not result in changes in the health status of mice or cause overt lung inflammation, indicating that the observed differences were due to the combined effects of bacterial infection and cigarette smoke exposure. Furthermore, the exacerbated inflammatory response was only observed when mice were inoculated with live bacteria; similar responses were observed between sham- and smoke-exposed mice following inoculation of inactivated bacteria. While this model lacks many of the clinical hallmarks of COPD, including chronic inflammation and airspace enlargement, it does demonstrate the complexity of the effects of cigarette smoke on innate antibacterial responses and the importance of using replicating infectious agents to model AECOPD.

Proteases and Bacterial Agents

Proteolytic enzymes, such as macrophage or neutrophil elastase, are important components of the antibacterial defense and are induced during immune responses. Proteolysis is further involved in cell recruitment and repair during an active infection or during the resolution of an infection [45]. Although proteolytic enzymes are essential for normal lung biology and defense, dysregulated expression may lead to tissue remodeling/destruction and disease [46]. This is of particular interest in conditions such as emphysema and COPD, where an imbalance between proteases and antiproteases has been implicated as an important factor leading to disease [47].

The most commonly used experimental model for the study of emphysema is achieved by instilment of elastase into the airways [48]. In this model, airspace enlargement has been shown to be dependent both on the enzyme's proteolytic activity and the inflammation that ensues as a result of the chemoattractive properties of degraded elastin fragments [49, 50]. There are few studies investigating the effect of elastase-induced emphysema on the antibacterial immune response as a model for AECOPD. However, recently, Inoue et al. [51] demonstrated that while control mice survived, mice instilled with elastase before inoculation with *S. pneumoniae* died in a dose-dependent manner, likely as a result both of excessive pulmonary and of systemic inflammation. These results indicate that under experimental conditions resembling emphysema, acute bacterial defense is impaired, resulting in excessive inflammation. Given that one of the clinical hallmarks of COPD is airspace enlargement, especially in late stage disease, this may contribute to the observed increases in bacterial infections and inflammation, as well as to the development of persistent bacterial colonization.

Inflammation has been suggested to be one of the driving forces underlying airspace enlargement. The combination of impaired antibacterial immune responses and a dysregulation of the protease/antiprotease balance may further contribute to tissue damage and disease progression. Along these lines, we observed greater proteolytic activity in the lungs of cigarette smoke- compared to sham-exposed mice following inoculation with *P. aeruginosa* [44]. The increased proteolytic burden was likely a reflection of the increased inflammation observed in these mice, as it was not observed in mice exposed to cigarette smoke only. Collectively, these findings indicate that cigarette smoke and bacterial infection have synergistic effects on airway damage, and that the damage may be mediated through effects on proteolytic activity. Such findings further underscore the importance of understanding the concurrent effect of cigarette smoke, protease regulation, and infection on mechanisms leading to AECOPD.

Bacterial Colonization

Bacterial colonization is believed to contribute to the pathogenesis of COPD by inducing a chronic inflammatory state thereby driving a progressive airway obstruction. Moreover, it has been shown to attenuate subsequent immune responses, which might perpetuate colonization or lead to a permanent inability to clear the pathogen [35]. Although an important hallmark of disease, bacterial colonization has been an understudied aspect of COPD in animal models, primarily because it is difficult to select relevant microbial species; pathogens which will colonize experimental animals often lack clinical applicability, while the clinically relevant pathogens do not often colonize animals.

When used to inoculate mice, pathogens and serotypes associated with AECOPD including nontypeable *H. influenzae* [52], *M. catarrhalis* [53], and *P. aeruginosa* [44] are cleared within a few days at most, and the elimination does not require the induction of major adaptive immune responses for clearance. *S. pneumoniae* has been shown to persist in some models [54], but not in others [55].

Several strategies have been pursued to prolong the persistence of these bacteria in the airways, including coupling them to agarose beads or another suitable medium [56], and the implantation of bacterium-coated tubing into the airways [57]. Although these methods prolong the pathogen's persistence in the airways, they do not obviate the fact that the bacterium is not actually colonizing the host. Alternatively, some murine respiratory pathogens, or human pathogens that have been adapted to murine systems, have been employed in models, which may shed light on AECOPD. Bacteria including *Mycoplasma pulmonis* [58], *Neisseria meningitidis* [59], *Bordetella parapertussis* and *B. bronchiseptica* [60, 61], and some isolates of *S. pneumoniae* [54] are all capable of colonizing the murine airway for significant periods of time under experimental conditions, and may therefore become useful as models for studying the impact of cigarette smoke on colonization and the consequences of colonization to respiratory host defense.

Alternatively, bacteria which would otherwise be noncolonizing can become colonizing agents in immunocompromised hosts [62]. The use of transgenic animals or immunodepleted hosts is an intriguing avenue for AECOPD research, as COPD patients may, in some respects, be viewed as immunocompromised. However, although useful for studying aspects of disease, this approach requires a detailed understanding not only of all the effects of a given gene deletion, but also of the immune status of COPD patients in order to correctly interpret data generated.

As previously mentioned, the bacteria most frequently associated with AECOPD is *H. influenzae*, an exclusively human bacterium found commensally in a large proportion of the population [12]. While modeling and comparing different strains of *H. influenzae* in animals are challenging because of the lack of infectivity in species other than humans [52], Chin et al. [53] recently tested the hypothesis that strains of *H. influenzae* associated with AECOPD induce more inflammation in mice than strains associated with asymptomatic colonization. Strains of *H. influenzae* associated with AECOPD or with colonization were inoculated into C57BL/6 mice in an in vivo model of airway infection. In this model, AECOPD-associated strains resulted in increased airway neutrophil recruitment that based on further in vitro experiments was likely mediated via increased induction of IL-8 and activation of the NF-κB and MAPK signaling pathways. These results indicate that strains of *H. influenzae* isolated during episodes of AECOPD may have different virulence compared to those isolated during stable disease, leading to augmented inflammation upon infection. Although this study did not address the consequences of cigarette smoke on inflammatory processes following inoculation with these bacterial strains, we speculate that cigarette smoke exposure would further exacerbate the inflammatory response.

While the antigens and virulence factors that lead to these differences have not been identified, studies such as these provide insight into mechanisms of airway inflammation observed during AECOPD. Moreover, they point to the importance of modeling different bacteria and environmental stimuli to understand the cellular and molecular mechanisms underlying both progression of stable COPD and exacerbations of this disease.

Modeling Viral Exacerbations

Several current studies have shown that as many as 40–60% of AECOPD are associated with respiratory virus infections [14, 63, 64]. In one study, a respiratory virus was detected in 56% of patients with COPD admitted to a hospital in Germany [65]. The major viruses that cause upper and lower respiratory tract infections and that are implicated in AECOPD include rhinovirus, respiratory syncytial virus, coronavirus, adenovirus, influenza A and B, and parainfluenza [14, 63, 64]. Among these, rhinovirus, the cause of common cold, is currently considered to be the most important trigger of COPD exacerbations [66]. As many as half of all colds during the peak fall cold season are estimated to be a result of rhinovirus infection [67]. Unfortunately, developing animal models to study rhinovirus infection has been met with some difficulties; rhinovirus is highly specific and does not recognize ICAM-1 from species other than humans [68]. Since this virus has significant implications in AECOPD, further work is warranted to develop animal models for the study of this virus.

Similar to the antibacterial host defense, cigarette smoke has been shown to compromise antiviral host defense mechanisms and increase the risk of respiratory viral infection. The incidence of influenza virus infection is increased in a smoking compared to a nonsmoking and nonimmunized population [69]. In addition, in vitro experiments from the 1980s suggested that the innate immune response, specifically that mediated by type I interferon is impaired upon exposure to cigarette smoke [70–72]. Taken together, these studies suggest that cigarette smoke affects innate antiviral immune defense mechanisms required to combat viral pathogens; hence, incorporating cigarette smoke into models of viral infections is essential for the understanding of virally induced AECOPD.

Adenoviruses have also been detected in AECOPD. DNA coding for the adenoviral E1A protein has been found in excess amounts in the lungs of patients with COPD when compared to controls matched for age, sex, and smoking history [73]. Adenoviral DNA has been shown to persist after acute infection, and viral E1A protein is expressed in lung epithelial cells long after the virus has stopped replicating [73, 74]. Retamales et al. [75] have shown that the excess inflammation observed in the lungs of smokers with severe emphysema is associated with increased numbers of alveolar epithelial cells expressing E1A. Importantly, Meshi et al. [76] showed a marked amplification of the cigarette smoke-induced inflammatory response and an increase in emphyse-matous lesion formation in guinea pigs with latent adenoviral infection when compared to noninfected controls. In mice, we have shown that cigarette smoke exposure was associated with attenuated T cell activation and decreased pro-duction of neutralizing antibodies following administration of a replication-deficient adenoviral construct [77]. This study may suggest that tobacco smoke-induced impairment of adaptive antiviral immunity contributes to the persistence of E1A protein in patients with COPD, although additional studies are required to provide direct evidence.

Recently, Behzad et al. [78] have determined the role of adenovirus infec-tion in inducing epithelial-mesenchymal transformation, an important feature of tissue remodeling. Remodeling processes contribute to (or cause) the obstruction of the small airways. This process is associated with a gradual decline in forced expiratory volume in 1 s, an important feature of COPD that is also linked to the progressive nature of the disease [3]. The authors demon-strated that adenovirus E1A is the key in epithelial-mesenchymal transforma-tion of primary lung epithelial cells derived from guinea pigs [78]. Taken together, these data suggest a role for adenoviruses in COPD pathogenesis and illustrate the need for further research of these models, specifically examining the underlying molecular implications cigarette smoke may have on virus-mediated AECOPD.

Influenza A and B viruses have also been shown to be important respira-tory viral infections leading to AECOPD. In one study of a nonimmunized pop-ulation between the ages of 60 and 90, rates of influenza infection in smokers were approximately 23% as compared to 6% in nonsmokers [69]. The relative contribution of this virus to AECOPD is thought to be less serious in regions that now offer influenza immunizations to patients with lung disease; however, it is still an important consideration as exacerbations may occur at times of influenza epidemics. In addition, recent data suggest that the level of protection by current influenza vaccines may be lower than what has been believed [79, 80]. Understanding the nature of viral infections in a healthy compared to ciga-rette smoke-exposed lung becomes important to understanding the role of viruses in AECOPD. Since cigarette smoking has been implicated in the changes observed in the COPD lung (as discussed above), superimposing a viral infection over the background of cigarette smoke adds a new element to animal modeling.

In a murine model of influenza virus infection, we have recently demon-strated that while viral clearance in sham- and smoke-exposed animals was similar, smoke exposure worsened inflammatory outcome and health status of animals [81]. The heightened inflammatory response observed in smoke-exposed animals was associated with increased expression of TNF-α, IL-6 and type I interferons in the airway, as well as with increased mortality [81]. In addition, we have shown that smoke exposure does not compromise the development of an influenza-specific memory response [81]. We postulate that the exaggerated airway inflammatory response to viral agents contributes to the inflammation observed in smokers and to the deterioration in clinical status associated with AECOPD.

Modeling Coinfection

Current disease models generally study interactions between a single pathogen and the infected/colonized host. While it is necessary to understand the basic science of disease, this 'single-agent:single-host' paradigm contrasts starkly with the situation in the clinic, where patients frequently present with multiple simultaneous infections, and where at the very least they present with a *history* of other infections. Moreover, as molecular biological techniques such as PCR have succeeded traditional clinical virological methods, it has become increasingly apparent that many COPD exacerbations are typically viral in nature [82]. Concomitantly, it has been suggested that viral infections may predispose a patient to bacterial infections, leading to exacerbations driven not by virus or bacteria alone, but by a bacterial-viral coinfection. Both from human studies and from experiments in animal models, it has become clear that the hosts' response to a pathogen is at least partially dictated by their immune system's history. In this regard, experimental models utilizing animal models and a single infection may not be predictive of an immunologically experienced COPD patient's response.

There is debate about the importance of heterologous viral-bacterial infections in the context of COPD; some studies have found correlations between these agents, while others have found no such relationships in patients with chronic bronchitis [83, 84]. However, these studies were performed without the aid of modern PCR methods, and therefore likely underestimated the true prevalence of viral-bacterial coinfections. More recently, a number of papers using PCR techniques have indicated that bacterial-viral coinfections are a major issue in clinical COPD management. Work by Papi et al. [85] found that approximately one quarter of all COPD exacerbations requiring hospitalization were of a mixed bacterial and viral etiology. Similarly, work by Cameron et al. [86] found that nearly one quarter of infectious exacerbations requiring ventilation were associated with simultaneous bacterial and viral infections. Wilkinson et al. [87] reported that exacerbations of putatively coinfectious etiology are marked by increased airflow limitation and symptomatology when compared to single-agent exacerbations. In addition to the epidemiological evidence, there is also ample experimental evidence to support the notion that infection with one agent predisposes to infection with another. Experimental influenza infection in humans increases the likelihood of natural colonization with *S. pneumoniae* [88]. Similarly, mice are more susceptible to *S. pneumoniae* following influenza A infection, and mount exacerbated inflammatory responses to the bacterium [89, 90].

Although the observation that two pathogens are more harmful to a host than one may not be surprising, investigating the mechanisms underlying this phenomenon is not trivial. In fact a number of mechanisms by which infection with one pathogen can modify the immune system's response to another pathogen have been proposed. Viral infection can effectively alter every stage of a bacterium's interaction with a host, by promoting bacterial attachment to epithelial cells [91, 92], impairing innate cell function [93, 94], and modulating the adaptive immune response [95]. Similarly, bacterial infections may alter preexisting antiviral immunity [96], and predispose individuals to subsequent viral infection [97].

Given that colonization of the airway is one of the defining characteristics of stable COPD, and that bacterial and viral exacerbations are a hallmark of acute exacerbation, modeling viral and bacterial infection over the background of colonization is likely to yield novel and exciting insight into AECOPD. Although complicated, such a model has recently been described by Seki et al. [98], in which *P. aeruginosa* colonization was achieved by implantation of bacterium-coated tubing into the airway, and animals were subsequently challenged with *S. pneumoniae* and influenza A virus. In this study, animals exposed to all three pathogens showed exacerbated cytokine responses, poor viral clearance, and reduced myeloperoxidase activity and lysozyme secretion compared to animals receiving only one or two of the pathogens. This and other developing models of colonization and coinfection must now be studied in detail, then overlaid on a background of other factors involved in the pathogenesis of COPD, such as cigarette smoke exposure, to realize their potential as models of AECOPD.

The current understanding is that infections are of critical importance to the pathogenesis of COPD. Given that a COPD patient's immune system is far from immunologically naïve, the need for animal models that examine heterologous infections in the context of COPD is apparent.

Conclusions

To date, there are no complete models of AECOPD. Instead, models of individual disease facets must be interpreted not only in the context of the specific aspect of human disease they mimic, but also in light of the effects other aspects may play. Ultimately, the most appropriate approach to modeling AECOPD is an inclusive one, in which we model various aspects of COPD, including cigarette smoke exposure, protease regulation, bacterial colonization, and mucus production, and then superimpose a relevant exacerbating stimulus. As such, we believe animal models will provide detailed mechanistic insight into the etiology of AECOPD and will prove invaluable in furthering our understanding of key events in disease pathogenesis.

References

1   Rodriguez-Roisin R: Toward a consensus definition for COPD exacerbations. Chest 2000;117: 398S–401S.
2   Pauwels RA, Buist AS, Calverley PM, Jenkins CR, Hurd SS: Global strategy for the diagnosis, management, and prevention of chronic obstructive pulmonary disease. NHLBI/WHO Global Initiative for Chronic Obstructive Lung Disease (GOLD) Workshop summary. Am J Respir Crit Care Med 2001;163:1256–1276.
3   Hogg JC: Pathophysiology of airflow limitation in chronic obstructive pulmonary disease. Lancet 2004;364:709–721.
4   Jeffery PK: Remodeling in asthma and chronic obstructive lung disease. Am J Respir Crit Care Med 2001;164:S28–S38.
5   Snider GL: Chronic obstructive pulmonary disease: a definition and implications of structural determinants of airflow obstruction for epidemiology. Am Rev Respir Dis 1989;140:S3–S8.
6   Wright JL, Churg A: Animal models of cigarette smoke-induced COPD. Chest 2002;122: 301S–306S.
7   Vlahos R, Bozinovski S, Gualano RC, Ernst M, Anderson GP: Modelling COPD in mice. Pulm Pharmacol Ther 2006;19:12–17.
8   Mallia P, Johnston SL: Mechanisms and experimental models of chronic obstructive pulmonary disease exacerbations. Proc Am Thorac Soc 2005;2:361–366, 371–372.
9   Mahadeva R, Shapiro SD: Chronic obstructive pulmonary disease * 3: experimental animal models of pulmonary emphysema. Thorax 2002;57:908–914.
10   Groneberg DA, Chung KF: Models of chronic obstructive pulmonary disease. Respir Res 2004;5:18.
11   Soler N, Torres A, Ewig S, Gonzalez J, Celis R, El-Ebiary M, Hernandez C, Rodriguez-Roisin R: Bronchial microbial patterns in severe exacerbations of chronic obstructive pulmonary disease (COPD) requiring mechanical ventilation. Am J Respir Crit Care Med 1998;157:1498–1505.
12   Sethi S, Murphy TF: Bacterial infection in chronic obstructive pulmonary disease in 2000: a state-of-the-art review. Clin Microbiol Rev 2001;14:336–363.
13   Wedzicha JA: Role of viruses in exacerbations of chronic obstructive pulmonary disease. Proc Am Thorac Soc 2004;1:115–120.
14   Papi A, Bellettato CM, Braccioni F, Romagnoli M, Casolari P, Caramori G, Fabbri LM, Johnston SL: Infections and airway inflammation in chronic obstructive pulmonary disease severe exacerbations. Am J Respir Crit Care Med 2006;173:1114–1121.
15   Eisner MD, Balmes J, Katz PP, Trupin L, Yelin EH, Blanc PD: Lifetime environmental tobacco smoke exposure and the risk of chronic obstructive pulmonary disease. Environ Health 2005;4:7.
16   Calverley PM: Reducing the frequency and severity of exacerbations of chronic obstructive pulmonary disease. Proc Am Thorac Soc 2004;1:121–124.
17   Patel IS, Seemungal TA, Wilks M, Lloyd-Owen SJ, Donaldson GC, Wedzicha JA: Relationship between bacterial colonisation and the frequency, character, and severity of COPD exacerbations. Thorax 2002;57:759–764.

18 Puchelle E, Zahm JM, Tournier JM, Coraux C: Airway epithelial repair, regeneration, and remodeling after injury in chronic obstructive pulmonary disease. Proc Am Thorac Soc 2006;3:726–733.

19 Murphy TF, Sethi S, Klingman KL, Brueggemann AB, Doern GV: Simultaneous respiratory tract colonization by multiple strains of nontypeable *Haemophilus influenzae* in chronic obstructive pulmonary disease: implications for antibiotic therapy. J Infect Dis 1999;180:404–409.

20 Hill AT, Campbell EJ, Hill SL, Bayley DL, Stockley RA: Association between airway bacterial load and markers of airway inflammation in patients with stable chronic bronchitis. Am J Med 2000;109:288–295.

21 Cabello H, Torres A, Celis R, El-Ebiary M, Puig de la Bellacasa J, Xaubet A, Gonzalez J, Agusti C, Soler N: Bacterial colonization of distal airways in healthy subjects and chronic lung disease: a bronchoscopic study. Eur Respir J 1997;10:1137–1144.

22 Sethi S, Evans N, Grant BJ, Murphy TF: New strains of bacteria and exacerbations of chronic obstructive pulmonary disease. N Engl J Med 2002;347:465–471.

23 Sethi S, Wrona C, Grant BJ, Murphy TF: Strain-specific immune response to *Haemophilus influenzae* in chronic obstructive pulmonary disease. Am J Respir Crit Care Med 2004;169:448–453.

24 Verra F, Escudier E, Lebargy F, Bernaudin JF, De Cremoux H, Bignon J: Ciliary abnormalities in bronchial epithelium of smokers, ex-smokers, and nonsmokers. Am J Respir Crit Care Med 1995;151:630–634.

25 Honda Y, Takahashi H, Kuroki Y, Akino T, Abe S: Decreased contents of surfactant proteins A and D in BAL fluids of healthy smokers. Chest 1996;109:1006–1009.

26 Subramaniam S, Whitsett JA, Hull W, Gairola CG: Alteration of pulmonary surfactant proteins in rats chronically exposed to cigarette smoke. Toxicol Appl Pharmacol 1996;140:274–280.

27 Twigg HL 3rd, Soliman DM, Spain BA: Impaired alveolar macrophage accessory cell function and reduced incidence of lymphocytic alveolitis in HIV-infected patients who smoke. Aids 1994;8:611–618.

28 Ortega E, Hueso F, Collazos ME, Pedrera MI, Barriga C, Rodriguez AB: Phagocytosis of latex beads by alveolar macrophages from mice exposed to cigarette smoke. Comp Immunol Microbiol Infect Dis 1992;15:137–142.

29 Ortega E, Barriga C, Rodriguez AB: Decline in the phagocytic function of alveolar macrophages from mice exposed to cigarette smoke. Comp Immunol Microbiol Infect Dis 1994;17:77–84.

30 Ohta T, Yamashita N, Maruyama M, Sugiyama E, Kobayashi M: Cigarette smoking decreases interleukin-8 secretion by human alveolar macrophages. Respir Med 1998;92:922–927.

31 McCrea KA, Ensor JE, Nall K, Bleecker ER, Hasday JD: Altered cytokine regulation in the lungs of cigarette smokers. Am J Respir Crit Care Med 1994;150:696–703.

32 King TE Jr, Savici D, Campbell PA: Phagocytosis and killing of *Listeria monocytogenes* by alveolar macrophages: smokers versus nonsmokers. J Infect Dis 1988;158:1309–1316.

33 Hodge S, Hodge G, Scicchitano R, Reynolds PN, Holmes M: Alveolar macrophages from subjects with chronic obstructive pulmonary disease are deficient in their ability to phagocytose apoptotic airway epithelial cells. Immunol Cell Biol 2003;81:289–296.

34 Hagiwara E, Takahashi KI, Okubo T, Ohno S, Ueda A, Aoki A, Odagiri S, Ishigatsubo Y: Cigarette smoking depletes cells spontaneously secreting Th(1) cytokines in the human airway. Cytokine 2001;14:121–126.

35 Berenson CS, Wrona CT, Grove LJ, Maloney J, Garlipp MA, Wallace PK, Stewart CC, Sethi S: Impaired alveolar macrophage response to haemophilus antigens in chronic obstructive lung disease. Am J Respir Crit Care Med 2006;174:31–40.

36 Berenson CS, Garlipp MA, Grove LJ, Maloney J, Sethi S: Impaired phagocytosis of nontypeable *Haemophilus influenzae* by human alveolar macrophages in chronic obstructive pulmonary disease. J Infect Dis 2006;194:1375–1384.

37 Burns AR, Hosford SP, Dunn LA, Walker DC, Hogg JC: Respiratory epithelial permeability after cigarette smoke exposure in guinea pigs. J Appl Physiol 1989;66:2109–2116.

38 Boucher RC, Johnson J, Inoue S, Hulbert W, Hogg JC: The effect of cigarette smoke on the permeability of guinea pig airways. Lab Invest 1980;43:94–100.

39 Meng QR, Gideon KM, Harbo SJ, Renne RA, Lee MK, Brys AM, Jones R: Gene expression profiling in lung tissues from mice exposed to cigarette smoke, lipopolysaccharide, or smoke plus lipopolysaccharide by inhalation. Inhal Toxicol 2006;18:555–568.

40 Thomas WR, Holt PG, Keast D: Cigarette smoke and phagocyte function: effect of chronic exposure in vivo and acute exposure in vitro. Infect Immun 1978;20:468–475.

41 Rylander R, Wold A, Haglind P: Nasal antibodies against gram-negative bacteria in cotton-mill workers. Int Arch Allergy Appl Immunol 1982;69:330–334.

42 Ozlu T, Cay M, Akbulut A, Yekeler H, Naziroglu M, Aksakal M: The facilitating effect of cigarette smoke on the colonization of instilled bacteria into the tracheal lumen in rats and the improving influence of supplementary vitamin E on this process. Respirology 1999;4:245–248.

43 Laurenzi GA, Guarneri JJ, Endriga RB, Carey JP: Clearance of bacteria by the lower respiratory tract. Science 1963;142:1572–1573.

44 Drannik AG, Pouladi MA, Robbins CS, Goncharova SI, Kianpour S, Stampfli MR: Impact of cigarette smoke on clearance and inflammation after *Pseudomonas aeruginosa* infection. Am J Respir Crit Care Med 2004;170:1164–1171.

45 Parks WC, Wilson CL, Lopez-Boado YS: Matrix metalloproteinases as modulators of inflammation and innate immunity. Nat Rev Immunol 2004;4:617–629.

46 Parks WC, Shapiro SD: Matrix metalloproteinases in lung biology. Respir Res 2001;2:10–19.

47 Hautamaki RD, Kobayashi DK, Senior RM, Shapiro SD: Requirement for macrophage elastase for cigarette smoke-induced emphysema in mice. Science 1997;277:2002–2004.

48 Janoff A, Sloan B, Weinbaum G, Damiano V, Sandhaus RA, Elias J, Kimbel P: Experimental emphysema induced with purified human neutrophil elastase: tissue localization of the instilled protease. Am Rev Respir Dis 1977;115:461–478.

49 Churg A, Wang RD, Tai H, Wang X, Xie C, Dai J, Shapiro SD, Wright JL: Macrophage metalloelastase mediates acute cigarette smoke-induced inflammation via tumor necrosis factor-alpha release. Am J Respir Crit Care Med 2003;167:1083–1089.

50 Houghton AM, Quintero PA, Perkins DL, Kobayashi DK, Kelley DG, Marconcini LA, Mecham RP, Senior RM, Shapiro SD: Elastin fragments drive disease progression in a murine model of emphysema. J Clin Invest 2006;116:753–759.

51 Inoue S, Nakamura H, Otake K, Saito H, Terashita K, Sato J, Takeda H, Tomoike H: Impaired pulmonary inflammatory responses are a prominent feature of streptococcal pneumonia in mice with experimental emphysema. Am J Respir Crit Care Med 2003;167:764–770.

52 Hansen EJ, Toews GB: Animal models for the study of noninvasive *Haemophilus influenzae* disease: pulmonary clearance systems. J Infect Dis 1992;165(suppl 1):S185–S187.

53 Chin CL, Manzel LJ, Lehman EE, Humlicek AL, Shi L, Starner TD, Denning GM, Murphy TF, Sethi S, Look DC: *Haemophilus influenzae* from patients with chronic obstructive pulmonary disease exacerbation induce more inflammation than colonizers. Am J Respir Crit Care Med 2005;172:85–91.

54 Kadioglu A, Taylor S, Iannelli F, Pozzi G, Mitchell TJ, Andrew PW: Upper and lower respiratory tract infection by *Streptococcus pneumoniae* is affected by pneumolysin deficiency and differences in capsule type. Infect Immun 2002;70:2886–2890.

55 Stark JM, Stark MA, Colasurdo GN, LeVine AM: Decreased bacterial clearance from the lungs of mice following primary respiratory syncytial virus infection. J Med Virol 2006;78:829–838.

56 van Heeckeren AM, Schluchter MD: Murine models of chronic *Pseudomonas aeruginosa* lung infection. Lab Anim 2002;36:291–312.

57 Yanagihara K, Tomono K, Sawai T, Hirakata Y, Kadota J, Koga H, Tashiro T, Kohno S: Effect of clarithromycin on lymphocytes in chronic respiratory *Pseudomonas aeruginosa* infection. Am J Respir Crit Care Med 1997;155:337–342.

58 Bakshi CS, Malik M, Carrico PM, Sellati TJ: T-bet deficiency facilitates airway colonization by *Mycoplasma pulmonis* in a murine model of asthma. J Immunol 2006;177:1786–1795.

59 Zhu D, Barniak V, Zhang Y, Green B, Zlotnick G: Intranasal immunization of mice with recombinant lipidated P2086 protein reduces nasal colonization of group B *Neisseria meningitidis*. Vaccine 2006;24:5420–5425.

60 Heininger U, Cotter PA, Fescemyer HW, Martinez de Tejada G, Yuk MH, Miller JF, Harvill ET: Comparative phenotypic analysis of the *Bordetella parapertussis* isolate chosen for genomic sequencing. Infect Immun 2002;70:3777–3784.

61 Harvill ET, Cotter PA, Miller JF: Pregenomic comparative analysis between *Bordetella bronchiseptica* RB50 and *Bordetella pertussis* tohama I in murine models of respiratory tract infection. Infect Immun 1999;67:6109–6118.

62 Kirimanjeswara GS, Mann PB, Harvill ET: Role of antibodies in immunity to Bordetella infections. Infect Immun 2003;71:1719–1724.
63 Qiu Y, Zhu J, Bandi V, Atmar RL, Hattotuwa K, Guntupalli KK, Jeffery PK: Biopsy neutrophilia, neutrophil chemokine and receptor gene expression in severe exacerbations of chronic obstructive pulmonary disease. Am J Respir Crit Care Med 2003;168:968–975.
64 Tan WC, Xiang X, Qiu D, Ng TP, Lam SF, Hegele RG: Epidemiology of respiratory viruses in patients hospitalized with near-fatal asthma, acute exacerbations of asthma, or chronic obstructive pulmonary disease. Am J Med 2003;115:272.
65 Rohde G, Wiethege A, Borg I, Kauth M, Bauer TT, Gillissen A, Bufe A, Schultze-Werninghaus G: Respiratory viruses in exacerbations of chronic obstructive pulmonary disease requiring hospitalisation: a case-control study. Thorax 2003;58:37–42.
66 Busse WW: Respiratory infections: their role in airway responsiveness and the pathogenesis of asthma. J Allergy Clin Immunol 1990;85:671–683.
67 Makela MJ, Puhakka T, Ruuskanen O, Leinonen M, Saikku P, Kimpimaki M, Blomqvist S, Hyypia T, Arstila P: Viruses and bacteria in the etiology of the common cold. J Clin Microbiol 1998;36: 539–542.
68 Bella J, Rossmann MG: Review: rhinoviruses and their ICAM receptors. J Struct Biol 1999; 128:69.
69 Nicholson KG, Kent J, Hammersley V: Influenza A among community-dwelling elderly persons in Leicestershire during winter 1993–4; cigarette smoking as a risk factor and the efficacy of influenza vaccination. Epidemiol Infect 1999;123:103–108.
70 Barnes MC, Streips UN, Sonnenfeld G: Effect of carcinogens and analogs on interferon induction. Oncology 1981;38:98–101.
71 Sonnenfeld G: Effect of sidestream tobacco smoke components on alpha/beta interferon production. Oncology 1983;40:52–56.
72 Sonnenfeld G, Hudgens RW: Effect of sidestream and mainstream smoke exposure on in vitro interferon-alpha/beta production by L-929 cells. Cancer Res 1986;46:2779–2783.
73 Matsuse T, Hayashi S, Kuwano K, Keunecke H, Jefferies WA, Hogg JC: Latent adenoviral infection in the pathogenesis of chronic airways obstruction. Am Rev Respir Dis 1992;146:177–184.
74 Elliott WM, Hayashi S, Hogg JC: Immunodetection of adenoviral E1A proteins in human lung tissue. Am J Respir Cell Mol Biol 1995;12:642–648.
75 Retamales I, Elliott WM, Meshi B, Coxson HO, Pare PD, Sciurba FC, Rogers RM, Hayashi S, Hogg JC: Amplification of inflammation in emphysema and its association with latent adenoviral infection. Am J Respir Crit Care Med 2001;164:469–473.
76 Meshi B, Vitalis TZ, Ionescu D, Elliott WM, Liu C, Wang XD, Hayashi S, Hogg JC: Emphysematous lung destruction by cigarette smoke. The effects of latent adenoviral infection on the lung inflammatory response. Am J Respir Cell Mol Biol 2002;26:52–57.
77 Robbins CS, Dawe DE, Goncharova SI, Pouladi MA, Drannik AG, Swirski FK, Cox G, Stampfli MR: Cigarette smoke decreases pulmonary dendritic cells and impacts antiviral immune responsiveness. Am J Respir Cell Mol Biol 2004;30:202–211.
78 Behzad AR, Morimoto K, Gosselink J, Green J, Hogg JC, Hayashi S: Induction of mesenchymal cell phenotypes in lung epithelial cells by adenovirus E1A. Eur Respir J 2006;28:1106–1116.
79 Jefferson T: Influenza vaccination: policy versus evidence. BMJ 2006;333:912–915.
80 Rivetti D, Jefferson T, Thomas R, Rudin M, Rivetti A, Di Pietrantonj C, Demicheli V: Vaccines for preventing influenza in the elderly. Cochrane Database Syst Rev 2006;3:CD004876.
81 Robbins CS, Bauer CMT, Vujicic N, Gaschler GJ, Lichty BD, Brown EG, Stampfli MR: Cigarette smoke impacts immune inflammatory responses to influenza in mice. Am J Respir Crit Care Med 2006;174:1342–1351.
82 Qiu Y, Zhu J, Bandi V, Atmar RL, Hattotuwa K, Guntupalli KK, Jeffery PK: Biopsy neutrophilia, neutrophil chemokine and receptor gene expression in severe exacerbations of chronic obstructive pulmonary disease. Am J Respir Crit Care Med 2003;168:968–975.
83 Gump DW, Phillips CA, Forsyth BR, McIntosh K, Lamborn KR, Stouch WH: Role of infection in chronic bronchitis. Am Rev Respir Dis 1976;113:465–474.
84 Smith CB, Golden C, Klauber MR, Kanner R, Renzetti A: Interactions between viruses and bacteria in patients with chronic bronchitis. J Infect Dis 1976;134:552–561.

85   Papi A, Bellettato CM, Braccioni F, Romagnoli M, Casolari P, Caramori G, Fabbri LM, Johnston SL: Infections and airway inflammation in chronic obstructive pulmonary disease severe exacerbations. Am J Respir Crit Care Med 2006;173:1114–1121.

86   Cameron RJ, de Wit D, Welsh TN, Ferguson J, Grissell TV, Rye PJ: Virus infection in exacerbations of chronic obstructive pulmonary disease requiring ventilation. Intensive Care Med 2006;32: 1022–1029.

87   Wilkinson TM, Hurst JR, Perera WR, Wilks M, Donaldson GC, Wedzicha JA: Effect of interactions between lower airway bacterial and rhinoviral infection in exacerbations of COPD. Chest 2006;129:317–324.

88   Wadowsky RM, Mietzner SM, Skoner DP, Doyle WJ, Fireman P: Effect of experimental influenza A virus infection on isolation of *Streptococcus pneumoniae* and other aerobic bacteria from the oropharynges of allergic and nonallergic adult subjects. Infect Immun 1995;63:1153–1157.

89   LeVine AM, Koeningsknecht V, Stark JM: Decreased pulmonary clearance of *S. pneumoniae* following influenza A infection in mice. J Virol Methods 2001;94:173–186.

90   Takase H, Nitanai H, Yamamura E, Otani T: Facilitated expansion of pneumococcal colonization from the nose to the lower respiratory tract in mice preinfected with influenza virus. Microbiol Immunol 1999;43:905–907.

91   Avadhanula V, Rodriguez CA, Devincenzo JP, Wang Y, Webby RJ, Ulett GC, Adderson EE: Respiratory viruses augment the adhesion of bacterial pathogens to respiratory epithelium in a viral species- and cell type-dependent manner. J Virol 2006;80:1629–1636.

92   Hament JM, Aerts PC, Fleer A, van Dijk H, Harmsen T, Kimpen JL, Wolfs TF: Direct binding of respiratory syncytial virus to pneumococci: a phenomenon that enhances both pneumococcal adherence to human epithelial cells and pneumococcal invasiveness in a murine model. Pediatr Res 2005;58:1198–1203.

93   Jakab GJ, Green GM: Defect in intracellular killing of *Staphylococcus aureus* within alveolar macrophages in Sendai virus-infected murine lungs. J Clin Invest 1976;57:1533–1539.

94   Jakab GJ, Green GM: The effect of Sendai virus infection on bactericidal and transport mechanisms of the murine lung. J Clin Invest 1972;51:1989–1998.

95   Qureshi MH, Garvy BA, Pomeroy C, Inayat MS, Oakley OR: A murine model of dual infection with cytomegalovirus and *Pneumocystis carinii*: effects of virus-induced immunomodulation on disease progression. Virus Res 2005;114:35–44.

96   Huang CC, Shah S, Nguyen P, Altman JD, Blackman MA: Bacterial superantigen exposure after resolution of influenza virus infection perturbs the virus-specific memory CD8(+)-T-cell repertoire. J Virol 2002;76:6852–6856.

97   Sajjan US, Jia Y, Newcomb DC, Bentley JK, Lukacs NW, LiPuma JJ, Hershenson MB: *H. influenzae* potentiates airway epithelial cell responses to rhinovirus by increasing ICAM-1 and TLR3 expression. FASEB J 2006;20:2121–2123.

98   Seki M, Higashiyama Y, Tomono K, Yanagihara K, Ohno H, Kaneko Y, Izumikawa K, Miyazaki Y, Hirakata Y, Mizuta Y, Tashiro T, Kohno S: Acute infection with influenza virus enhances susceptibility to fatal pneumonia following *Streptococcus pneumoniae* infection in mice with chronic pulmonary colonization with *Pseudomonas aeruginosa*. Clin Exp Immunol 2004;137:35–40.

Martin R. Stämpfli
Department of Pathology and Molecular Medicine
Centre for Gene Therapeutics, McMaster University
Hamilton, ON L8N 3Z5 (Canada)
Tel. +1 905 525 9140, ext. 22493, Fax +1 905 522 6759, E-Mail stampfli@mcmaster.ca

Author Index

Bauer, C.M.T. 126
Busse, W.W. 12

Caramori, G. 101
Cates, E.C. 42
Churg, A. 113
Contoli, M. 101

Evans, M.Y. 21

Fattouh, R. 42

Gaschler, G.J. 126
Gauvreau, G.M. 21

Johnson, J.R. 42
Johnston, S.L. 101
Jordana, M. 42

Kharitonov, S.A. 83

Lindell, D. 68
Llop-Guevara, A. 42
Lukacs, N.W. 68

Mallia, P. 101

O'Byrne, P.M. 1

Papadopoulos, N.G. 33
Papi, A. 101

Schaller, M. 68
Singh, A.M. 12
Sjöbring, U. VIII, 83
Smit, J. 68
Stämpfli, M.R. 126

Taylor, J.D. VIII

Wright, J.L. 113

Xatzipsalti, M. 33

Zavitz, C.C.J. 126

Allergen inhalation challenge (AIC)
  drug efficacy testing 29, 30
  early-phase asthmatic response
    features 22
    mechanisms 23
  isolated early responders and dual
    responders 27–29
  late-phase asthmatic response
    features 22
    mechanisms 24–26
  nonspecific airway hyperresponsiveness
    26, 27
  precautions 22
  technique 22
Anti-monocyte chemoattractant protein-1,
  lipopolysaccharide challenge test
  findings 95
Asthma exacerbations
  allergen inhalation challenge model, *see*
    Allergen inhalation challenge
  chronic obstructive pulmonary disease
    exacerbation comparison 8, 9
  clinical manifestations 4, 5
  definitions 2, 3
  epidemiology 6–8
  mortality 5
  respiratory syncytial virus, *see*
    Respiratory syncytial virus
  rhinovirus models, *see* Rhinovirus

Bacteria, *see* Chronic obstructive
  pulmonary disease exacerbations;
  Lipopolysaccharide
Bronchitis, animal models 122, 123

Chronic obstructive pulmonary disease
  (COPD) exacerbations
  animal models
    bacterial infection models
      colonization 128, 131–133
      protease mediation 130, 131
      smoke synergy 129, 130
    bronchitis 122, 123
    challenges 126
    coinfection models 135, 136
    prospects 136, 137
    smoking models, *see* Emphysema;
      Pulmonary hypertension; Small
      airway resistance
    virus infection models 133–135
  asthma exacerbation comparison
    8, 9
  clinical manifestations 5, 6
  definitions 3, 4, 127
  disease stage and outcomes 114
  epidemiology 8, 101
  frequency and severity determinants
    127, 128
  lipopolysaccharide induction, *see*
    Lipopolysaccharide
  mortality 6
  virus infections, *see also* Rhinovirus
    detection 103
    immune response 102, 103
    mechanism of exacerbation
      cellular mechanisms 104, 105
      inflammatory mechanisms 104
      oxidative stress 105
    susceptibility 105, 106

Cilomilast, lipopolysaccharide challenge
test findings 94, 95
Corticosteroids, lipopolysaccharide
challenge test findings 95

Dendritic cells (DC), subsets in respiratory
syncytial virus induced asthma
exacerbation 73, 74
Dust mite, *see* House dust mite

Elastase, knockout mouse studies of chronic
obstructive pulmonary disease 116
Emphysema, smoke-induced, animal
models
anatomic patterns 114–116
immune response 120
intervention studies
anti-inflammatory therapy 119
antioxidant protection 118, 119
metalloprotease inhibition 117, 118
serine protease inhibitors 116, 117
mouse strain differences 119, 120
Endotoxin, *see* Lipopolysaccharide

House dust mite (HDM)
allergy
allergens 46, 47, 52–55
models
acute exposure 47–49
chronic asthma 49–51
exposure considerations 56–59
sampling of allergens 56–58
prevalence 46
biology 46
chitinases 54
extracts 52–55
lipopolysaccharide 55
proteases 52–54

Intercellular adhesion molecule-1 (ICAM-1)
expression in rhinovirus-induced asthma
exacerbation 16, 17, 34, 35
virus-induced chronic obstructive
pulmonary disease exacerbation role
104, 106
Interferon-β (IFN-β), expression in
rhinovirus-induced asthma exacerbation
17, 39

Interleukin-6 (IL-6), virus-induced chronic
obstructive pulmonary disease
exacerbation role 104
Interleukin-10 (IL-10), expression in
rhinovirus-induced asthma exacerbation
15, 16

Lipid A, see Lipopolysaccharide
Lipopolysaccharide (LPS)
binding and Toll-like receptor-4 signaling
85, 86
challenge testing
bronchial blood flow 91
dosing 89
drug testing
anti-monocyte chemoattractant
protein-1 95
cilomilast 94, 95
corticosteroids 95
salmeterol 94
exhaled breath condensate 91
exhaled nitric oxide measurement 90
exhaled temperature measurement 91
healthy individual findings
bronchoalveolar lavage 93
respiratory symptoms and lung
function 92, 93
sputum biomarkers 93
systemic biomarkers 92
systemic effects 91, 92
intranasal challenge 95, 96
intravenous administration 96
sampling timing 89, 90
smoker findings 94
sputum biomarkers 90
technique 88, 89
chronic obstructive pulmonary disease
bacteria role in exacerbations 88
exacerbation challenge findings 94
lower respiratory tract colonization 87,
88
oral cavity sources 88
function 84
house dust mite extracts 55
lipid A toxicity 85
occupational exposure 86
structure 84
tobacco smoke 84, 86, 87

Matrix metalloprotease-12, knockout mouse
    studies of chronic obstructive pulmonary
    disease 117
Monocyte chemoattractant protein-1
    antibody, *see* Anti-monocyte
    chemoattractant protein-1

Nitric oxide, exhalation measurement in
    lipopolysaccharide challenge 90, 91
Nrf2, knockout mouse studies of chronic
    obstructive pulmonary disease 118, 119
Nuclear factor-κB (NF-κB), virus-induced
    chronic obstructive pulmonary disease
    exacerbation response 105

Ovalbumin (OVA), asthma models 44, 45
Oxidative stress
    smoke-induced chronic obstructive
        pulmonary disease role 118, 119
    virus-induced chronic obstructive
        pulmonary disease exacerbation role 105

Pulmonary hypertension, smoke-induced,
    animal models 121, 122

RANTES (regulated on activation, normal T
    cell expressed, and secleted), expression
    in rhinovirus asthma exacerbation
    15–17, 39
Respiratory syncytial virus (RSV)
    antibody therapy 69
    asthma exacerbation
        dendritic cell subsets 73, 74
        epidemiology 68, 69
        T cell recruitment 75, 76
    biology 69, 70
    chemokine response 74, 75
    economic impact 69
    epidemics 69
    innate immune response 70, 71, 73
Rhinovirus
    asthma exacerbation
        cell models
            A549 cells 36
            BEAS-2B cells 35, 36
            fetal lung fibroblasts 36, 37
            human airway smooth muscle
                cells 37

human nasal epithelial cells 38
            peripheral blood nuclear cells 37, 38
        epidemiology 13, 33
        experimental infection in humans
            allergic sensitization synergism
                17, 18
            immunopathogenesis 14–17
            prospects for study 19
        microbiology 13, 14
        mouse models 34, 35
        overview 12, 13
    chronic obstructive pulmonary disease
        exacerbation, human model 106–109
Roflumilast, chronic obstructive pulmonary
    disease protection, mouse model 119

Salmeterol, lipopolysaccharide challenge
    test findings 94
Scientific modeling, overview 43, 44, 59
Small airway resistance (SAR), smoke-
    induced, animal models 121, 123
Smoking
    chronic obstructive pulmonary disease
        models, *see* Emphysema; Small airway
        resistance
    immune response 120
    lipopolysaccharides
        challenge test findings in smokers 94
        tobacco smoke 84, 86, 87
Statins, chronic obstructive pulmonary
    disease protection, mouse model 119

Tachycardia, asthma exacerbations 4
T cell
    recruitment in respiratory syncytial virus
        induced asthma exacerbation 75, 76
    virus-induced chronic obstructive
        pulmonary disease exacerbation
        response 105
Tobacco, *see* Smoking
Toll-like receptors (TLRs)
    lipopolysaccharide binding and Toll-like
        receptor-4 signaling 85, 86
    respiratory syncytial virus innate immune
        response 70, 71, 73
Tumor necrosis factor-α receptor, knockout
    mouse studies of chronic obstructive
    pulmonary disease 119